Materials Forming, Machining and Tribology

Series Editor

J. Paulo Davim, Department of Mechanical Engineering, University of Aveiro,
Aveiro, Portugal

This series fosters information exchange and discussion on all aspects of materials forming, machining and tribology. This series focuses on materials forming and machining processes, namely, metal casting, rolling, forging, extrusion, drawing, sheet metal forming, microforming, hydroforming, thermoforming, incremental forming, joining, powder metallurgy and ceramics processing, shaping processes for plastics/composites, traditional machining (turning, drilling, miling, broaching, etc.), non-traditional machining (EDM, ECM, USM, LAM, etc.), grinding and others abrasive processes, hard part machining, high speed machining, high efficiency machining, micro and nanomachining, among others. The formability and machinability of all materials will be considered, including metals, polymers, ceramics, composites, biomaterials, nanomaterials, special materials, etc. The series covers the full range of tribological aspects such as surface integrity, friction and wear, lubrication and multiscale tribology including biomedical systems and manufacturing processes. It also covers modelling and optimization techniques applied in materials forming, machining and tribology. Contributions to this book series are welcome on all subjects of "green" materials forming, machining and tribology. To submit a proposal or request further information, please contact Dr. Mayra Castro, Publishing Editor Applied Sciences, via mayra.castro@springer.com or Professor J. Paulo Davim, Book Series Editor, via pdavim@ua.pt

More information about this series at http://www.springer.com/series/11181

Sunil Pathak

Editor

Intelligent Manufacturing

Springer

Editor
Sunil Pathak
Faculty of Manufacturing and Mechatronics
Engineering Technology
University Malaysia Pahang
Pekan, Malaysia

ISSN 2195-0911 ISSN 2195-092X (electronic)
Materials Forming, Machining and Tribology
ISBN 978-3-030-50314-7 ISBN 978-3-030-50312-3 (eBook)
https://doi.org/10.1007/978-3-030-50312-3

This Springer imprint is published by the registered company Springer Nature Switzerland AG
The registered company address is: Gewerbestrasse 11, 6330 Cham, Switzerland

Preface

The 4th industrial revolution urges all sectors, including manufacturing, to develop and adopt intelligent methods and techniques in order to stay competitive in the global economy. The manufacturing sector is one of the most significant contributors in the world economy and always busy promoting research, development, and innovations to meet the accelerated demand for productivity, quality, and sustainability. Intelligent manufacturing processes and techniques have always been helpful to attain that. Therefore, it is important for the engineers, managers, researchers, and professors working in the manufacturing field to understand the development, implementation and effectiveness of various intelligent manufacturing methods and techniques.

This book facilitates them by providing fundamental understanding, basic knowledge, and advanced research insights on intelligent manufacturing methods and techniques for a wide range of conventional and advanced type manufacturing processes.

This book consists of eight chapters on intelligent manufacturing. Chapter 1 presents a detailed review of the usefulness of optimization techniques in electro-discharge machining. Chapter 2 sheds lights on the optimization of machining parameters for material removal rate and machining time while cutting Inconel 600 with tungsten carbide textured tools and explains the effectiveness of TOPSIS, GRA, and MOORA for machinability enhancement of Inconel 600-type superalloys. Chapter 3 presents a Kurtosis analysis of tool drilling geometries and cutting conditions for deep twist drilling process improvement and provides details on the optimization of parameters using design of experiments techniques. Chapter 4 highlights the development of Ti50Ni50-xCox (x = 1 and 5 at %) shape memory alloy and investigation of input process parameters of wire spark discharge machining, and the study highlights the usefulness of GRA and entropy measurement methods. Chapter 5 describes a thorough study on the application potential of fuzzy embedded TOPSIS approach to solve MCDM-based problems. Whereas, Chap. 6 sheds highlights on multi-criteria decision making through soft computing, and evolutionary techniques provide a comprehensive review of various intelligent techniques for optimization of manufacturing processes. Chapter 7 presents an advanced study on the application of multi-criteria decision-making techniques in the optimization of mechano-tribological properties of copper metal matrix sic-graphite-reinforced composites. The book ends with Chap. 8 highlighting the hybrid approach for prediction and modelling of abrasive water jet machining parameter on Al-NiTi composites, where the combinations of advanced optimization techniques such as SVM-entropy model, GRA-AVM-entropy modelling.

I sincerely acknowledge Springer for this opportunity and their professional support. I am also thankful to all chapter contributors for their availability and valuable contributions.

Kuantan, Malaysia   Sunil Pathak

Contents

Nomenclature

ANN	Artificial neural network
AP	Gas pressure
B	Bed speed
C	Capacitance
CP	Concentration of powder
DF	Duty factor
DP	Dielectric pressure
DT	Discharge time
EDM	Electro-discharge machining
EW	Electrode wear
EWR	Electrode wear rate
F	Feed rate
G	Gain size of powder
GA	Genetic algorithm
GC	Gap current
GRA	Grey relational analysis
I	Discharge current
IP	Peak current
KERF	Cutting width
M	Machining depth
MD	Machining diameter
MID	Microscopic depth
MRR	Material removal rate
OC	Radial overcut
P	Polarity
PW	Pulse width
R	Resistance
RC	Radial clearance of shield at the bottom
REWR	Relative electrode wear rate
RLT	Recast layer thickness

RS	Residual stress
RSM	Response surface methodology
S	Electrode speed
SCD	Surface crack density
SF	Flushing rate
SF	Sparking frequency
SG	Spark gap
SH	Surface hardness
SHF	Shape factor
SR	Surface roughness
SUF	Surface finish
SV	Servo voltage
T	Tool electrode lift time
TA	Tool area
TM	Total machining time
T_{off}	Pulse off time
T_{on}	Pulse on time
TS	Tool shape
TT	Taguchi technique
TWR	Tool wear rate
V	Open/gap voltage
W	Wire speed
WLT	White layer thickness
WT	Wire tension
WWR	Wire wear ratio

Chapter 1
A Study on Optimization Techniques of Electro Discharge Machining

Shatarupa Biswas, Yogesh Singh, and Manidipto Mukherjee

Abstract The technique of optimization is a mathematical method for calculating the maximum and minimum value of a function f(x), which is subject to certain constraints. Optimization is the expression, under the given circumstance, of obtaining the best mix of outcomes. No specific methods are available to deal with all numerical and mathematical issues. So different types of techniques for dealing with different kinds of issues have been introduced. The optimization technique is most widely used in Electro Discharge Machining (EDM) to find the best combination of input parameters (such as current, voltage, pulse on time, pulse off time, etc.) for the desired output parameters (such as material removal rate, surface roughness, tool wear rate, etc.). This method usually only applies to conductive materials (such as silver, copper, aluminium, brass, bronze, etc.). Electricity can flow through the conductive materials for conductivity. Once dielectric fluid (acts as a semiconductor and flushing agent to clean and remove eroded debris from the region of the spark gap, such as kerosene, EDM oil, and deionized water) flows, the workpiece is released metal ions, and the tool is released electrons. Therefore, a spark is formed between the tool and workpiece and induced temperature of 8000–12000 °C, and due to this spark energy, the material is removed from the substrate. The maximum MRR (Material removal rate), minimum TWR (Tool wear rate) and SR (Surface roughness) are critical for the optimization technique. In this study, different types of optimization techniques are defined for EDM machining purposes based on their uses.

S. Biswas · Y. Singh (✉)
Mechanical Engineering Department, National Institute of Technology, Silchar,
Assam 788010, India
e-mail: yogeshsingh15@gmail.com

S. Biswas
e-mail: supriticu@gmail.com

M. Mukherjee
CAMM, CSIR-Central Mechanical Engineering Research Institute, Durgapur,
West Bengal 713209, India
e-mail: m.mukherjee.ju@gmail.com

© Springer Nature Switzerland AG 2021
S. Pathak (ed.), *Intelligent Manufacturing*, Materials Forming, Machining
and Tribology, https://doi.org/10.1007/978-3-030-50312-3_1

Keywords EDM (electro discharge Machining) · ANN (artificial neural network) · TT (Taguchi technique) · GRA (grey relational Analysis) · GA (genetic algorithm) · RSM (response surface methodology)

1.1 Introduction

In the 1770s, an English physicist Joseph Priestley discovered EDM machining techniques. It was developed monetarily in the mid-1980s. EDM is a non-traditional material cutting method mainly used for comparatively hard metals that would be very difficult for machining using the current conventional process such as shaping, broaching, boring, turning, etc. No physical contact is needed between a tool and workpiece, and they are connected to the DC power supply as the main features of the EDM machining process. Using spark, which is produced between the tool and the workpiece, the workpiece is eroded. Several metal alloys such as Titanium, Hastelloy, Kovar, and Inconel were introduced to this process. EDM machining relies on process parameters such as current, voltage, pulse on time, pulse off time etc. [1, 2] and dielectric fluids such as kerosene, deionized water, EDM oil etc. to boost the different types of material performance parameters, namely SR, SH, TWR, MRR, RLT, microstructures. It also demonstrates the use of various types of optimization techniques to accomplish various tasks.

1.1.1 Type of EDM

Generally, three types of EDM are used in engineering applications:

1. Die sinking EDM
2. Wire cut EDM
3. Micro EDM.

1.1.1.1 Die Sinking EDM

EDM's Die sinking version is a state-of-the-art manufacturing technology commonly used in several industrial sectors [1]. Several-research on this type of EDM [2–5] has already been carried out. Die sinking EDM is also known as cavity type EDM. The electrode and workpiece are immersed in the dielectric fluid when the machining process is underway. Both the workpiece and the tool are linked to a sufficient power supply. The dielectric fluids break down into positive and negative ions while the power supply is on. After that, ions are released from the workpiece and the electrode and mixed together. Spark is developed for this purpose and hits on the workpiece to remove metals by erosion. Typically, in one second, one hundred thousand sparks may be generated. The process parameter (such as current, voltage, pulse on time,

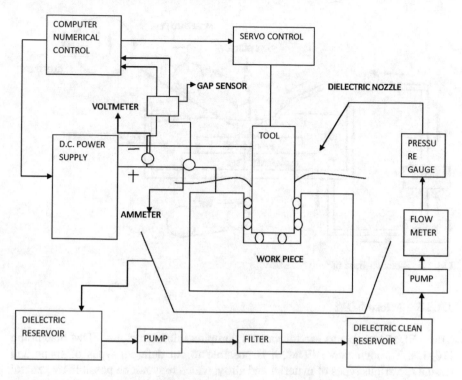

Fig. 1.1 Block diagram of die sinking EDM

pulse off time, etc.) must be regulated within a duty cycle to control the machining process. The schematic representation of the method is shown in Fig. 1.1.

1.1.1.2   Wire Cut EDM

Typically used for cutting purposes, this type of EDM [6–9] also called spark EDM. In principle, this type of EDM, a thin wire, is used as a tool. The wire is primarily made of brass or graphite. The diameter of the wire will range from 0.1 to 0.3 mm and for machining, the plate thickness is essentially 300 mm. This EDM is usually used when low residual stresses are needed, as high cutting forces are not necessary to remove the material. Wire EDM process makes machining burr-free and provided the workpiece is electrically conductive. The schematic representation of the method is shown in Fig. 1.2.

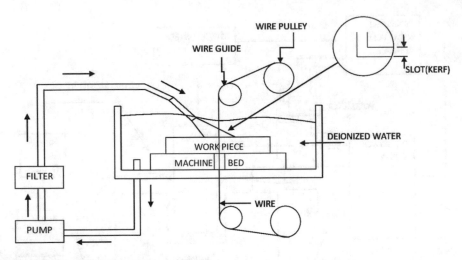

Fig. 1.2 Block diagram of wire cut EDM

1.1.1.3 Micro EDM

Micro EDM has able to use for various complex microstructures EDM machining [10–13]. Through micro EDM, it is possible to cut deferent types of shape and design of various types of material and alloy, which may not be possible by general EDM machining. Stress-free micro-sized cavities of requiring shapes on conducting and semiconducting materials are possible by using micro EDM. In micro EDM workpiece (anode) and tool (cathode) separated by the dielectric fluid are supplied with pulsed voltage. The workpiece and tool are brought closer until the dielectric in the media breaks down and allow current to pass through it, which appears as sparks. Micro EDM process can be separated into four categories based on the tool types and tool kinematics. They are μ-die-sinking EDM, μ-EDM drilling, μ-milling EDM & μ-wire EDM. The schematic representation of the method is shown in Fig. 1.3.

1.1.2 Important Parameters in EDM

EDM machining is generally depended on process and performance parameters. Deferent types of EDM machining parameters are shown in Fig. 1.4. For EDM machining, some process parameters have to choose based on various types of materials. Because the performance parameters of the materials are depended on the progress parameters has chosen during EDM machining.

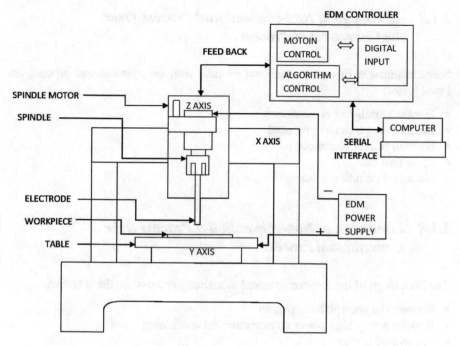

Fig. 1.3 Block diagram of micro EDM

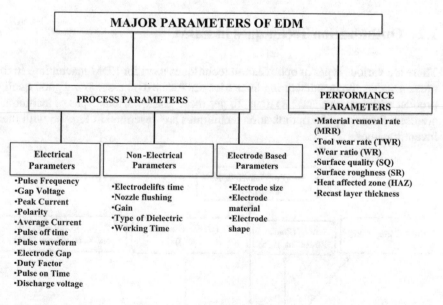

Fig. 1.4 Major parameters of EDM

1.1.3 Advantage of Non-conventional Process Over the Conventional Process

Some advantages of non-conventional process over the conventional process are listed below:

- Hard materials can be machined
- Intricate shape can be produced
- No chip formation create
- More tool life
- No sound pollution occur.

1.1.4 Limitation of Non-conventional Process Over Conventional Process

The limitations of the non-conventional machining process are listed below:

- Require the good skilled operator
- Require a very high power consumption for machining
- Set up cost is high
- The material removal process is slow.

1.2 Optimization Techniques in EDM

There are various types of optimization techniques used for EDM machining. In the present status, the manufacturing industries are focused on assembling good quality product with less cost and less time. To get these goals, various types of techniques are used. Some popular optimization techniques are described in Fig. 1.5 with their inventor names.

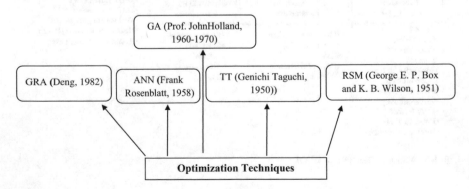

Fig. 1.5 Different Optimization techniques use for EDM machining

1.2.1   Grey Relational Analysis (GRA)

Deng [14] develops the grey relational technique in 1982. GRA is used for solving the inner relationship among the several characteristics. In GRA hypothesis incorporates three kinds of frameworks first dark, which shows no data in this framework, second white which shows all data in this framework and third dim framework, which shows erroneous data. The GRA is a productive method, which needs a bit of limited data to assess the conduct of a vulnerability framework and limited information issue. If the scope of the grouping is enormous, the elements are effaceable. The main advantages of these techniques are the output is based on the input data, calculation steps easy. So it is used mostly in the engineering and management area. In this technique, standardization of the trial results accomplished for MRR, TWR, and SR. The three states of GRA are listed in the below [14].

1. Lower value is the better

$$x_i^*(k) = \frac{X_i(K) - \min X_i(K)}{\max X_i(K) - \min X_i(K)} \qquad (1.1)$$

2. Higher value is the better

$$x_i^*(k) = \frac{\max X_i(K) - X_i(K)}{\max X_i(K) - \min X_i(K)} \qquad (1.2)$$

3. Nominal value is the best

$$x_i^*(k) = \frac{1 - |X_i(K) - X_o b(K)|}{\max X_i(K) - X_o b(K)} \qquad (1.3)$$

The standardization is taken by the ensuing conditions. Where k = 1, I = 1, 2n, 2, y, p; $x_i^*(k)$ is the standardized estimation of the kth component in the ith arrangement, $x_0 b(k)$ is an estimation of the kth quality element, max $x_i^*(k)$ is the biggest estimation of $x_i(k)$, and min $x_i^*(k)$ is the littlest estimation of $x_i(k)$, n is the number of examinations and p is the quantity of value attributes. The working flowchart of GRA has appeared in Fig. 1.6.

The uses of GRA technique in different materials for improving the performance of EDM machining is shown in Table 1.1.

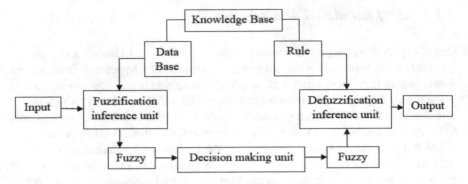

Fig. 1.6 Flowchart diagram for grey relational analysis

1.2.2 Artificial Neural Network (ANN)

For EDM machining, one of the most common optimization techniques is ANN. It is a data handling framework that is normal with organic neural systems. Counterfeit neural systems have been created as speculation of arithmetic models of human insight or neural science, in light of certain doubts. Like Signals are disregarded between neurons association joins. Every association link has a related weight, which, in many neural nets, duplicates the sign transmitted and each neuron applies an activation function to its net input to determine its output signal. The structure of this system contains the system's information, quantities of shrouded layers with various neurons in every layer and an outside layer with neurons associated with yield. The main advantages of the techniques are work with incomplete data, store capability, distributed memory, the parallel capability of processing and disadvantages are does not shows the optimum values of parameters, the time duration of network. A multilayer discernment with one concealed layer has appeared in Fig. 1.7, and the working flowchart of ANN has appeared in Fig. 1.8.

The uses of the ANN technique in different materials for improving the performance of EDM machining is shown in Table 1.2.

1.2.3 Genetic Algorithm (GA)

The GA methodology is influenced by the theory of evolution of Darwin [14]. The algorithm begins with a record of experiments. Any natural parallelism and inclination data is not needed for this system when looking through the space of the program. It is currently an effective, scalable strategy for advancement. The GA specification was reviewed in EDM by a few specialists. The main advantage of this approach is that if there is no statistical analysis, it is efficient and but the method is prolonged.

Table 1.1 Various research on EDM machining based on the GRA technique

Year	Author	Material	Electrode	Types of EDM	Input parameters	Output parameters	Remarks
2010	Kao et al. [15]	Ti–6Al–4V	Copper	Die sink	I, V, T_{on}, DF	MRR, SR, EWR	It showed that the EWR, MRR, and SR are improved by 12%,15% and 19% respectively
	Reza et al. [16]	AISI 304	Copper	–	P, T_{on}, I, V, M, MD, DP	MRR, SR, EWR	The output parameters are 16.36% improved
	Jung and Kwon [17]	SS 304	WC (Tungsten carbide)	Wire cut	V, C, R, F, S	EW	It was found that the V and C are established to be the most leading parameter
2012	Moghaddam et al. [18]	40CrMnMoS86	Copper	Die sink	IP, DF, V, Ton, T_{off}	MRR, SR, TWR	TT & GRA is quite well-organized in determining optimal for EDM process parameters
	Rajesh and Anand [19]	–	Copper	Die sink	I, V, DP, SG, T_{on}, T_{off}	MRR, SR	Working current is the most influencing factors for the EDM machining
2012	Singh [20]	6061 Al/Al2O3p/20P Al	Copper	Die sink	I, T_{on}, DF, V, T	MRR, SR, TWR	GRA approach can be applied effectively for different tasks in which carrying out measures are controlled by many procedure parameters at various quality solicitations

(continued)

Table 1.1 (continued)

Year	Author	Material	Electrode	Types of EDM	Input parameters	Output parameters	Remarks
2013	Raghuraman et al. [21]	MS IS 2026	Copper	Die sink	IP, Ton, T_{off}	MRR, TWR, SR	GRA technique is useful for optimization of SR, TWR, and MRR
2013	Gopalakannan et al. [22]	Al 6063	Copper	Die sink	I, V, T_{on}, T_{off}	MRR, EWR, SR	The grey relation enhances of single EDM quality which is 27.71%
2013	Ganachari et al. [23]	AISI D3 steel	Copper	Die sink	CP, T_{on}, DF, V, IP	SR	The goal of this testing is to find the sequence of input parameters to obtain the superior surface finish
2018	Upadhyay et al. [24]	Inconel 601	Copper	Die sink	IP, T_{on}, V, DF, DP	MRR, EWR, SR, SCD	The best compromise solution appear as IP 6A, T_{on} 200 μs, V 50v, DF 75%, DP 0.2 bar & T_{on} is the most significance parameter

Fig. 1.7 Bubble diagram for artificial neural network

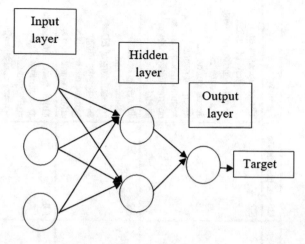

Fig. 1.8 Flowchart diagram for artificial neural network

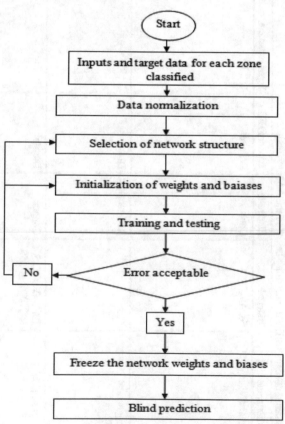

Table 1.2 Various research on EDM machining based on the AAN technique

Year	Author name	Material	Electrode	Types of EDM	Input parameters	Output parameters	Remarks
2001	Tsai and Wang [25]	Fe, Al	Copper	Die sink	DT, IP	SR	The comparison of expectations of surface completion for different work materials dependent on six divergent neural systems models and a neuron-fluffy system model
2004	Fenggou and Dayong [26]	S136	Red cooper	Die sink	IP, PW	SR	The programmed assurance and streamlining of EDM sinking preparing parameters by ANN is effective and relevant
2008	Rao et al. [27]	Ti6Al4V, HE15, 15CDV6, M-250	Copper	Die sink	IP, V	MRR	Multi observation neural system models are utilizing neuro arrangements bundle
2009	Pradhan et al. [28]	AISI D2	Copper	Die sink	I, T_{on}, DF	SR	Two distinctive ANN models of back-proliferation neural system and outspread premise work neural system are significant to anticipate SR

(continued)

Table 1.2 (continued)

Year	Author name	Material	Electrode	Types of EDM	Input parameters	Output parameters	Remarks
2010	Pradhan and Biswas [29]	AISI D2	Copper	Die sink	I, T_{on}, DF, V	MRR, TWR, OC	The investigation shows that the release current is the most prevailing component
	Mahdavinejad [30]	SiC	Copper	Die sink	I, T_{on}, T_{off}	MRR, SR	ANN by using back proliferation calculation is utilized to modelled the procedure. MRR & SR both are streamlined as destinations by utilizing NSGA-II
2010	Yahya et al. [31]	Steel	Copper	Die sink	GC, T_{on}, T_{off} SF	MRR	The ANN model is fit for anticipating the MRR with a low rate
2012	Bharti et al. [32]	Inconel 718	Copper	Die sink	SHF, T_{on}, I, DF, V, SF, T	MRR, SR	The normal rate contrast among trial and ANN's anticipated worth is 4 and 4.67 for MRR and SR separately
2013	Tzeng and Chen [33]	JIS SKD 61 steel	Copper	Die sink	I, V, T_{on}, T_{off}	MRR, REWR, SR	The back proliferation neural system GA shows the better expectation

(continued)

Table 1.2 (continued)

Year	Author name	Material	Electrode	Types of EDM	Input parameters	Output parameters	Remarks
2016	Gosavi and Gaikwad [34]	EN 31	copper	Die sink	I, T_{on}, T_{off}, V	MRR	The outcomes acquired by numerical examination and trial strategies have been looked at. It tends to be presumed that the numerical technique gives a sensibly precise estimation of reactions

Fig. 1.9 Flowchart diagram
for genetic algorithm

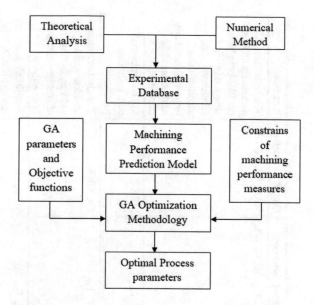

Figure 1.9 demonstrates the working theory or steps. The uses of the GA technique
in different materials for improving the performance of EDM machining are shown
in Table 1.3.

1.2.4 Taguchi Technique (TT)

This approach is one of the wide-spread and effective processes of optimization that
help to improve the quality of existing products, processes and cost reduction with
minimal machining. This approach accomplishes by rendering the execution of the
process "heartless" to factors such as tools, hardware production, workmanship and
working conditions. The main advantage of this approach shows the best combination
of input parameters that are mostly influenced by the output parameters and the main
drawback does not show the highest effective input parameters. There are three
classifications of execution attributes in the examination of the Signal to noise (S/N)
proportion as pursues [14].

1. Larger value is the better characteristics

$$\frac{S}{N} = -10\log\left(\frac{1}{n}\sum_{i=1}^{n}\frac{1}{Yi^2}\right) \qquad (1.4)$$

2. Smaller value is the better characteristics

$$\frac{S}{N} = -10\log\left(\frac{1}{n}\sum_{i=1}^{n}Yi^2\right) \qquad (1.5)$$

Table 1.3 Various research on EDM machining based on the GA technique

Year	Author	Material	Electrode	Types of EDM	Input parameters	Output parameters	Remarks
2007	Mahapatra and Patnaik [35]	D2 tool steel	Zinc coated copper wire	Wire cut	I, T_{on}, SF, W, WT, F	MRR, SR, KERF	The procedure parameters of Wire cut EDM can be changed in accordance with accomplishing improved machining exhibitions at the same time
2008	Gao et al. [36]	C40 steel	Copper	Die sink	I, T_{on}, T_{off}	MRR	MRR is enhanced with optimized parameters
2010	Maji and Pratihar [37]	Mild steel	Copper	Die sink	IP, T_{on}, DF	SR, MRR	The ideal outcomes are identified as good and Pareto-ideal front of arrangements had been acquired
2011	Mahdavinajad [30]	SiC	Copper	Die sink	I, T_{on}, T_{off}	MRR, SR	A multi advancement strategy NSGA-II is applied and at last Pareto-ideal arrangements of MRR and SR are obtained
2011	Yahya et al. [31]	steel	Copper	Die sink	GC, T_{on}, T_{off}, SF	MRR	ANN model is equipped for anticipating the MRR with low rate expectation blunder

(continued)

Table 1.3 (continued)

Year	Author	Material	Electrode	Types of EDM	Input parameters	Output parameters	Remarks
2012	Bharti et al. [32]	Inconel 718	Copper	Die sink	SF, T_{off}, I, DF, V, SF, T	MRR, Surface finish	The normal rate contrast among exploratory and ANN's anticipated worth is 4 & 4.67 for MRR and SR separately
	Padhee et al. [38]	EN 31	Copper	Die sink	IP, T_{on}, DF, CP	MRR, Surface finish	Enhancing Both MRR and SR, NSGA II is embraced to acquire the Pareto ideal arrangement
	Rajesh and Anand [19]	Al alloy, Grade HE9	Copper	Die sink	I, V, DP, SG, T_{on}, T_{off}	MRR, SR	The technique is utilized for augmentation of MRR & minimization of SR
2013	Tzeng and Chen [33]	SKD61	Copper	Die sink	I, V, T_{on}, T_{off}	MRR, EWR, SR	GA has better expectation and compliance result than the RSM technique

(continued)

Table 1.3 (continued)

Year	Author	Material	Electrode	Types of EDM	Input parameters	Output parameters	Remarks
2017	Mahapatra and Sahoo [35]	Ti-6Al-4V	Brass	Wire cut	T_{on}, T_{off}, F, SV	KERF, MRR, SR, WWR	Small scale auxiliary examination was done for both the wires and the work-piece at the advanced settings of the reaction acquired for both metal and combi wire anodes

3. Nominal value is the better characteristics

$$\frac{S}{N} = -10 \log \left(\frac{y}{Sy^2} \right) \tag{1.6}$$

Where Yi is the tentatively watched worth and n is rehashed number of each examination y is the normal of watched information, and Sy^2 is the difference of Yi for each kind of attributes, with the above S/N change, the higher the S/N proportion, the better is the outcome. The flowchart using this technique is shown in Fig. 1.10.

The uses of TT technique in different materials for improving the performance of EDM machining are shown in Table 1.4.

Fig. 1.10 Flowchart diagram for Taguchi technique

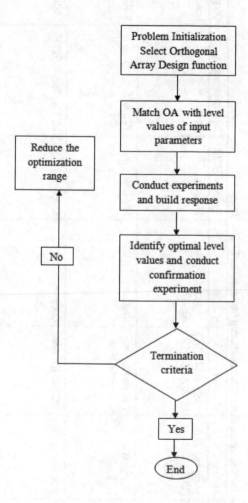

Table 1.4 Various research on EDM machining based on the TT technique

Year	Author	Material	Electrode	Types of EDM	Input parameters	Output parameters	Remarks
2001	Wang and Tsai [39]	AISI, AISI D2, EK2, AISI H13	Copper, Graphite, Ag–W	Die sink	IP, T_{on}, P	MRR & TWR	A semi observational model of the material erosion rate on the job is set up by utilizing dimensional analysis
2004	George et al. [40]	Carbon-carbon composite plate	Copper	Die sink	T_{on}, I, V	MRR, TWR	The EDM procedure variable influencing the MRR &TWR as indicated by their relative importance
2005	Kansal et al. [41]	HCHCr die steel	Copper	Die sink	IP, T_{on}, T_{off}, V, T, SF	MRR, TWR, SR	Expansion of a fitting measure of the graphite powder with the dielectric liquid caused discernible improvement in MRR and decreased in wear and SR
2006	Lin et al. [42]	HSS of grade SKH 57	Copper	Die sink	P, IP, V, T_{on}	MRR, EWR, SR	MRR expanded with the IP. As the beat length broadened, the MRR at first expanded to a top at around 100 µs and dropped

(continued)

Table 1.4 (continued)

Year	Author	Material	Electrode	Types of EDM	Input parameters	Output parameters	Remarks
	Kansal et al. [43]	H11 die steel	Copper	Die sink	IP, T_{on}, DF, CP	MRR, TWR, SR	It is observed that grouping of included silicon powder in dielectric liquid and pinnacle flow are the most persuasive parameters for TWR, MRR, and SR
	Chattopadhyay et al. [44]	EN-8	Copper	Rotary dry	IP, T_{on}, S	MRR, EWR, SR	The primary impact factor influencing the selected presentation measure factors, for example, MRR, EWR and SR
	Govindan and Joshi [45]	SS 304	Copper, Aluminium	Rotary dry	I, V, T_{off}, DP, S, RC	MRR, TWR	At low release energies, single-release in dry electric release machining could give bigger MRR and hole sweep contrasted with dielectric EDM
2011	Nipanikar and Ghewae [46]	Inconel 718	Copper	Die sink	T_{on}, IP, DF, V	MRR, EWR, OC	The IP altogether influences the MRR and OC; Ton primarily influences the EWR

(continued)

Table 1.4 (continued)

Year	Author	Material	Electrode	Types of EDM	Input parameters	Output parameters	Remarks
2012	Syed and Palaniyandi [47]	W300 die-steel	Copper	Die sink	IP, T_{on}, P, CP	MRR, EWR, SR, WLT	Inclusion of Al powder in distilled water is improved the MRR, great SF & lower the WLT
2013	Vhatkar et al. [48]	EN-31	Copper	Die sink	IP, T_{on}, T_{off}, V, CP	MRR, SR	It is observed that with the inclusion of the powders in the dielectric fluids, MRR is improved, SR is decreased
	Amit et al. [49]	EN-5	Copper	Die sink	IP, T_{on}, DP	MRR	TT method is utilized for the advancement of reaction factors
2013	Bergaley and Sharma [50]	HCHcr D3 steel	Copper	Die sink	V, I, T_{on}, T_{off} DP, S	MRR, EWR	MRR and EWR is effected by input parameters
2014	Goyal et al. [51]	AISI 1045	Copper	Die sink	I, V, T_{on}, DF, CP	MRR, SR	The grain size of aluminium powder and convergence of aluminium powder blended in with EDM oil which impacts the MRR and SR

(continued)

Table 1.4 (continued)

Year	Author	Material	Electrode	Types of EDM	Input parameters	Output parameters	Remarks
2015	Nagaraja et al. [52]	Al_2O_3		Wire cut	T_{on}, T_{off}, F	MRR, SR	It is observed from the result F is the most crucial machining parameter for SR
2016	Singh et al. [53]	Ti-6Al-4V	Copper	Die sink	T_{on}, T_{off}, I	SR	It is found that the set of parameters for SR is Ton (24 µs) T_{off} (9 µs) and I is (5amp)
2017	Kumar et al. [54]	Inconel 718	Brass	Wire cut	WT, W, I, T_{on}	MRR, SR	The ideal condition is found that Ton 30 µs for MRR and other parameters are WT 16 N, W 305 mm/s, I 20A and Ton 22 µs
2017	Mohanty et al. [55]	D2 steel	Tool is produced by direct metal laser sintering	Die sink	IP, T_{on}, T_{off}, DP	TWR, MRR, SR	Utility-based TT is the significant for evaluation of the perfect parameter setting

1.2.5 Response Surface Method (RSM)

The technique is investigated by Box and Draper in 1951 [14]. It is a statistical and mathematical technique. A second-degree polynomial model is used in the RSM technique. Because it is so easy to estimate. It is mainly used for making a model and doing an analysis of problems in which a response of interest is influenced by many variables. The development of an adequate functional relationship between responses of interest. The model is quadratic. The first-order model is

$$y = \beta_0 + \sum_{i=1}^{k} \beta_i X_i + \xi i \tag{1.7}$$

If there is a curvature in the reaction surface, then a higher degree polynomial should be used.

$$Y = \beta_0 + \sum_{i=1}^{k} \beta_i X_i + \sum_{i=1}^{k} \beta_{ii} X_i^2 + \sum_{i,j=1 i=j}^{k} \beta_{ij} X_i X_j + \xi \tag{1.8}$$

where the error can be identified in reaction Y. X_i is the linear input factors, X_i^2 and $X_i X_j$ are the squares and association terms. The second-order coefficients are β_0, β_i, β_j and β_{ii}. It is determined by the second-order model and least square method. The RSM includes a series of experiments for measurement of the response and make a mathematical model of the second-order response surface with the best fittings. Finding the optimal parameters the Minitab Software was used. The main advantage of this technique shows the pure error between actual values and predicted values. The flowchart using this technique is shown in Fig. 1.11.

The uses of RSM technique in different materials for improving the performance of EDM machining is shown in Table 1.5.

1.3 Comparison of All Techniques

From the last few decades, plenty of research has been performed under the different optimization techniques for the EDM machining process. As in the previous section, different optimization techniques for the EDM process are discussed, and the comparison between them is presented in Table 1.6. The comparison between the different optimization techniques are observed, and it is found that the performance improvement of EDM machining is better using GRA and RSM optimization techniques, as shown in Table 1.6.

Fig. 1.11 Flowchart diagram for response surface method

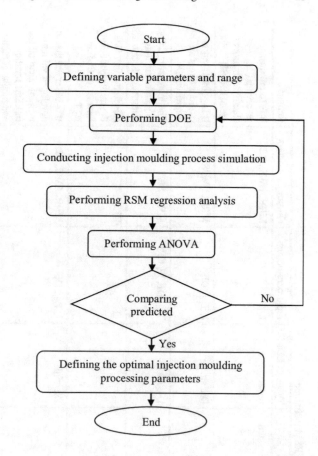

1.4 Conclusions

It is evident from the overall study, that to work with EDM is very important to work with optimization techniques. Through the optimization processes, a satisfactory result can be obtained. During EDM machining, an appropriate optimization technique can save both costs and time. Different types of techniques of optimization and use are discussed. It is shown that for all events or conditions, a single optimization strategy may not be fruitful. The main advantages of the GRA approach are the output parameter dependent on the input data, and the measurement steps are relatively easy and quick compared to the other methods.

On the other hand, the main advantages of the ANN technique will operate with incomplete data. The main drawbacks of this process are that it does not represent the optimal performance parameter values. The main advantage of the GA technique is that if there is no mathematical analysis, it is useful, but compared to the other methods, the calculation process is lengthy. TT is the most common method of optimization used to evaluate optimum input and set parameters for EDM. This

Table 1.5 Various research on EDM machining based on the RSM technique

Year	Author	Material	Electrode	Types of EDM	Input parameters	Output parameters	Remarks
1994	Soni and Chakraverti [56]	Ti6A14V	Copper-tungsten	Rotary	I, S	MRR, EWR, SR	It is found that revolving the electrode improves the MRR, SR
2004	Puertas et al. [57]	94WC–6Co	Copper	Die sink	I, T_{on}, DF	MRR, EW, SR	It is pointed out that in order to get an improved SR
2005	Kansal et al. [58]	EN 31	Copper	Die sink	T_{on}, DF, IP, CP	MRR, SR	The increasing CP in the dielectric fluid improves SR & MRR
2007	Chiang et al. [59]	Ferritic SG cast iron	Copper	–	G	RLT	It is found that G is the most significant factors for finding RLT
2007	Luis and Puertas [60]	B_4C, SiSiC, WC-Co	Copper	Die sink	I, T_{on}, DF	MRR, EWR, SR	It is observed that to developed values of technological tables in EDM for the conductive ceramic materials
2008	Kuppan et al. [61]	Inconel 718	Copper	Rotary	IP, T_{on}, DF, S	MRR, SR	MRR is more affected by IP, DF and S
2008	Chiang [62]	Al_2O_3 + TiC	Copper	Die sink	I, T_{on}, DF, V	MRR, EWR, SR	The principal components for MRR are I and DF

(continued)

Table 1.5 (continued)

Year	Author	Material	Electrode	Types of EDM	Input parameters	Output parameters	Remarks
2009	Patel et al. [63]	Al_2O_3/SiCw/TiC ceramic composite	Copper	Die sink	I, T_{on}, DF, V	SR	It is found that SR is improved by using RSM technique
	Sohani et al. [64]	EN8	Copper	Die sink	IP, T_{on}, T_{off}, TA, TS	MRR, TWR	The best apparatus shape for higher MRR and lower TWR is round
	Saha and Choudhury [65]	EN32 mild steel	Copper	Die sink	V, I, T_{on}, DF, AP, S	MRR, TWR, SR	It is found that I, DF, AP, and S are mostly affected by MRR and SR
2009	Habib [66]	Al/SiC	Copper	Die sink	T_{on}, IP, V, CP	MRR, EWR, SR	In this study, output parameters are improved
	Pradhan and Biswas [67]	AISI D2	Copper	Die sink	I, T_{on}, T_{off}	SR	It is discovered that I, T_{on}, T_{off} and not many of their collaborations have a critical impact on the SR
2010	Iqbal and Khan [68]	AISI 304	Copper	EDM milling	V, S, F	MRR, EWR, SR	In this study, V and S are the most critical machining parameters which effect the MRR, EWR, and SR

(continued)

Table 1.5 (continued)

Year	Author	Material	Electrode	Types of EDM	Input parameters	Output parameters	Remarks
2012	Padhee et al. [38]	EN 31	Copper	Die sink	CP, T_{on}, DF, IP	MRR, SUF	Scientific models for the forecast of MRR and SR through the information of four process factors are created utilizing RSM
	Rajesh and Anand [19]	Al Alloy with Grade HE9	Copper	Die sink	I, V, DP, SG, T_{on}, T_{off}	MRR, SUF	Exact models for MRR and SR are made by conduction an organized test
2012	Jabbaripour et al. [69]	Ti–6Al–4V	–	–	IP, T_{on}, V	MRR, TWR	In this study, when IP & T_{on} increased then improved output parameters
2013	Shandilya et al. [70]	Al 6061	Brass	Wire cut	SV, T_{on}, T_{off}, F	SR	It is observed that SR value is decreased for using RSM technique
	Assarzadeh and Ghoreishi [71]	Tungsten carbide cobalt composite	–	–	I, T_{on}, DF, V	MRR, TWR, SR	Here using RSM, improved MRR, TWR, and SR.
2013	Rajendran et al. [72]	T90Mn2W50Cr45 tool steel	Copper	Die sink	T_{on}, T_{off}, I	EW, RLT	In this study 'I' is directly proportional with the RLT and SCD.

(continued)

Table 1.5 (continued)

Year	Author	Material	Electrode	Types of EDM	Input parameters	Output parameters	Remarks
	Ayesha et al. [73]	C1023	Graphite	–	I, T_{on}, SV	EW, TM	The best parameters for low EW and low deterioration time is those that join low I, high T_{on} and low SV
	Tzeng and Chen [33]	SKD61	Copper	Die sink	I, V, T_{on}, T_{off}	MRR, EWR, SR	The higher release vitality with the expansion of I and T_{on} improved MRR.
	Syed and Kuppan [74]	W300 die steel	Copper	Die sink	IP, T_{on}, CP	WLT	Optical microscopy results show that low WLT is 17.14 μm
2014	Gupta and Jain [75]	Brass	Brass	Wire cut	V, T_{on}, T_{off}, F	SR	It is discovered that V, T_{on}, T_{off}, F were critical variables influencing the normal and greatest SR of the WEDMed small scale gears
2016	Talla et al. [76]	Inconel 625	Copper	Die sink	I	SR, SCD, WLT, MID, and residual stress	It is found that the least SR value was acquired at 6 g/l powder fixation
2016	Shandilya et al. [77]	SiCp/6061	Brass	Wire cut	V, T_{on}, T_{off}, F	MRR, KERF	It is observed that the SV and T_{off}main factors of MRR

(continued)

Table 1.5 (continued)

Year	Author	Material	Electrode	Types of EDM	Input parameters	Output parameters	Remarks
2017	Bhaumk and Maity [78]	AISI 304	Tungsten carbide	Die sink	IP, T_{on}, V DF, CP	MRR, SR, TWR	It is found that the optimal parametric mix of I: 4A, T_{on}: 150 μs, V: 65 V, DF: 65% and CP: 10 g/l

Table 1.6 Comparison of all optimization techniques based on output results

Optimization techniques	Output parameters		
	MRR	SR	KERF
GRA	Improved MRR value (Ti-6Al-4V) is 12% compare to actual value. The actual value is 2.25 mg/min and improved value is 3.20 mg/min [15]	Improved SR value (Ti-6Al-4V) is 19% compare to actual value. The actual value is 2.20 μm and improved value is 2.78 μm [15]	–
ANN	Shows low % error (Ti6Al4V, HE15, 15CDV6, M-250) with compare to the actual value, the actual value is 22.41 mm^3/min, and the predicted value is 24.92 mm^3/min. Shows %error is 2.27% [27]	Shows low % error (SKD 61) with compare to the actual value, the actual value is 4.52 μm, and the predicted value is 4.49 μm. Shows %error is 3.00% [33]	–
GA	Improved actual MRR value (Hot Die Steel) is 9.84–9.95 mm^3/min [79]	Improved actual value SR value (Hot Die Steel) is 4.23–4.15 mm [79]	–
TT	Improved MRR value (EN 5) with the effect of IP. The value is 0.41 mg/s [49]	When T_{on} decreased then SR decreased (D2 Tool Steel). The minimum value is 3.5019 μm [80]	After using the TT (D2 Tool Steel), the decreased value is 0.3291 mm [80]
RSM	After using the RSM the maximum MRR value (Tungsten carbide cobalt composite) is 0.287 mm^3/min [71]	After using the RSM the minimum SR value (Tungsten carbide cobalt composite) is 2.710 μm [71]	After using the RSM KERF width (SiCp/6061) decreased by 7.69%. The value decreased 0.211–0.204 mm [77]

Note
GRA: Input parameters: I is an essential parameter for MRR
TT: Input parameters: I&DF, V are the essential parameter for MRR. TT gives a good value between predicted and experimental values
RSM: Input parameters: I, T_{on}, T_{off} are affecting parameters of SR and T_{on} are significant factors for MRR

technique shows the best combination of input parameters mainly influenced by output parameters and this approach is relatively easy and less complicated but does not show the highest practical input value for EDM. For RSM technique, the pure error for the output parameter between the actual values and the expected values is shown. The methodology of optimization should be used based on the needs and specifications. Some of the technique of optimization is good for a specific condition, and a considerable error can be made for different cases with the same technique.

References

1. K. Gupta, M.K. Gupta, Developments in non-conventional machining for sustainable production—a state of art review. Proc. IMechE, Part C J. Mech. Eng. Sci. (Sage) **233**(12), 4213–4232 (2019)
2. B.K. Sahua, S. Datta, On electro-discharge machining of Inconel 718 super alloys : an experimental investigation. Mater. Today Proc. **5**, 4861–4869 (2018)
3. N. Sharma, K. Gupta, J.P. Davim, On wire spark erosion machining induced surface integrity of Ni55.8Ti shape memory alloys. Arch. Civ. Mech. Eng. (Elsevier) **19**(3), 680–693 (2019)
4. P.K. Patowari, P. Saha, P.K. Mishra, An experimental investigation of surface modification of C-40 steel using W–Cu powder metallurgy sintered compact tools in EDM. Int. J. Adv. Manuf. Technol. (2015)
5. M. Rahang, P.K. Patowari, Application of masking technique in EDM for generation of rectangular shaped pattern. Int. J. Precis. Technol. **5**(2), 140–156 (2015)
6. K. Gupta, N.K. Jain, Comparative study of Wire-EDM and hobbing for manufacturing high-quality miniature gears comparative study of wire-EDM and hobbing for manufacturing high-quality Miniature gears. Mater. Manuf. Process. (October), 37–41 (2014)
7. K. Gupta, N.K. Jain, Deviations in geometry of miniature gears fabricated by wire electrical discharge machining, in: *Proceedings of International Mechanical Engineering Congress & Exposition (IMECE 2013) of ASME*, V010T11A047, Nov. 13–21, 2013, San Diego, California, USA (2013)
8. K. Mohapatra, S. Sahoo, A multi objective optimization of gear cutting in WEDM of Inconel 718 using TOPSIS method. Decis. Sci. Lett. **7**, 157–170 (2018)
9. A. Kumar, H. Mishra, K. Vivekananda, K.P. Maity, Multi-objective optimization of wire electrical discharge machining process parameterson Inconel 718. Mater. Today Proc. **4**(2), 2137–2146 (2017)
10. C. Wang, Comparison of micro-EDM characteristics of Inconel 706 between EDM Oil and an Al powder-mixed dielectric. Adv. Mater. Sci. Eng. **2019** (2019)
11. M.Y. Ali, A. Sabur, A. Banu, A. Maleque, E.Y.T. Adesta, Micro electro discharge machining of nonconductive ceramic. Int. J. Eng. Mater. Manuf. **3**, 65–72 (2018)
12. K.P. Somashekhar, S. Panda, J. Mathew, N. Ramachandran, Numerical simulation of micro-EDM model with multi-spark. Int. J. Adv. Manuf. Technol. (2013)
13. L. Raju, S.S. Hiremath, A state-of-the-art Review on micro electro-discharge machining. Procedia Technol. **25**(Rarest), 1281–1288 (2016)
14. S.K. Choudhary, R.S. Jadoun, A. Kumar, Latest research trend of optimization techniques in electric discharge machining (EDM): review article. Int. J. Res. Eng. Adv. Technol. **2**(3), 1–29 (2014)
15. J.Y. Kao, C.C. Tsao, S.S. Wang, C.Y. Hsu, Optimization of the EDM parameters on machining Ti-6Al-4V with multiple quality characteristics. Int. J. Adv. Manuf. Technol. **47**(1–4), 395–402 (2010)
16. M.S. Reza, M. Hamdi, M.A. Azmir, Optimization of EDM injection flushing type control parameters using grey relational analysis on AISI304 Stainless steel workpiece, in *National Conference in Mechanical Engineering Research and Postgraduate Students (1st NCMER 2010)*, no. May, pp. 564–571 (2010)
17. J.H. Jung, W.T. Kwon, Optimization of EDM process for multiple performance characteristics using Taguchi method and Grey relational analysis. J. Mech. Sci. Technol. **24**(5), 1083–1090 (2010)
18. M.A.M. Azadi Moghaddam, F. Kolahan, Application of grey relational analysis and simulated annealing algorithm for modeling and optimization of EDM parameters on 40CrMnMoS86 hot worked steel, in *20th Annual International Conference on Mechanical Engineering*, vol. 53, no. 9, pp. 1689–1699 (2012)
19. R. Rajesh, M. Dev Anand, The optimization of the electro-discharge machining process using response surface methodology and genetic algorithms. Procedia Eng. **38**, 3941–3950 (2012)

20. S. Singh, Optimization of machining characteristics in electric discharge machining of 6061Al/Al2O3p/20P composites by grey relational analysis. Int. J. Adv. Manuf. Technol. **63**(9–12), 1191–1202 (2012)
21. S.S. Raghuraman, K. Thiruppathi, T. Panneerselvam, Optimization of EDM parameters using Taguchi method and grey relational analysis for mild steel is 2026. Int. J. Innov. Res. Sci. Eng. Technol. **2**(7), 3095–3104
22. S. Gopalakannan, T. Senthilvelan, S. Ranganathan, Statistical optimization of EDM parameters on machining of aluminum Hybrid Metal Matrix composite by applying Taguchi based Grey analysis. J. Sci. Ind. Res. (India) **72**(6), 358–365 (2013)
23. V.S. Ganachari, M.V. Kavade, S.S. Mohite, Effect of mixture of Al and Sic powder on surface roughness in PMEDM using Taguchi method with GRA optimization. Int. J. Adv. Eng. Res. Stud., 04–07 (2013)
24. C. Upadhyay, Rahul, S. Datta, S.S. Mahapatra, B.B. Biswal, An experimental investigation on electro discharge machining of Inconel 601. Int. J. Ind. Syst. Eng. **29**(2), 223–251 (2018)
25. K.M. Tsai, P.J. Wang, Predictions on surface finish in electrical discharge machining based upon neural network models. Int. J. Mach. Tools Manuf. **41**(10), 1385–1403 (2001)
26. C. Fenggou, Y. Dayong, The study of high efficiency and intelligent optimization system in EDM sinking process. J. Mater. Process. Technol. **149**(1–3), 83–87 (2004)
27. G.K.M. Rao, G.R. Janardhan, D.H. Rao, M.S. Rao, Development of hybrid model and optimization of metal removal rate in electric discharge machining using artificial neural networks and genetic algorithm. ARPN J. Eng. Appl. Sci. **3**(1), 19–30 (2008)
28. M.K. Pradhan, R. Das, C.K. Biswas, Comparisons of neural network models on surface roughness in electrical discharge machining. J. Eng. Manuf. **223**, 801–808 (2009)
29. M. Kumar, P. Chandan, K. Biswas, Neuro-fuzzy and neural network-based prediction of various responses in electrical discharge machining of AISI D2 steel. Int. J. Adv. Manuf. Technol., 591–610 (2010)
30. R.A. MahdaviNejad, Modeling and optimization of electrical discharge machining of SiC parameters, using neural network and non-dominating sorting genetic algorithm (NSGA II). Mater. Sci. Appl. **02**(06), 669–675 (2011)
31. A. Yhya, T. Andromeda, A. Baharom, A.A. Rahim, N. Mahmud, Material removal rate prediction of electrical discharge machining process using artificial neural network. J. Mech. Eng. Autom. **1**, 298–302 (2011)
32. P.S. Bharti, S. Maheshwari, C. Sharma, Multi-objective optimization of electric-discharge machining process using controlled elitist NSGA-II. J. Mech. Sci. Technol. **26**(6), 1875–1883 (2012)
33. C.J. Tzeng, R.Y. Chen, Optimization of electric discharge machining process using the response surface methodology and genetic algorithm approach. Int. J. Precis. Eng. Manuf. **14**(5), 709–717 (2013)
34. A.B.G. Abhinandan, A. Gosavi, Predicting optimized EDM machining parameter through thermo mechanical analysis. IOSR J. Mech. Civ. Eng. **13**(3), 71–85 (2016)
35. S.S. Mahapatra, A. Patnaik, Optimization of wire electrical discharge machining (WEDM) process parameters using Taguchi method. Int. J. Adv. Manuf. Technol., 911–925 (2007)
36. Q. Gao, Q.H. Zhang, S.P. Su, J.H. Zhang, Parameter optimization model in electrical discharge machining process. J. Zhejiang Univ. Sci. A **9**(1), 104–108 (2008)
37. K. Maji, D.K. Pratihar, Modeling of electrical discharge machining process using conventional regression analysis and genetic algorithms. J. Mater. Eng. Perform. **20**(7), 1121–1127 (2011)
38. S. Padhee, N. Nayak, S.K. Panda, P.R. Dhal, S.S. Mahapatra, Multi-objective parametric optimization of powder mixed electro-discharge machining using response surface methodology and non-dominated sorting genetic algorithm. Sadhana—Acad. Proc. Eng. Sci. **37**(2), 223–240 (2012)
39. P.J. Wang, K.M. Tsai, Semi-empirical model on work removal and tool wear in electrical discharge machining. J. Mater. Process. Technol. **114**(1), 1–17 (2001)
40. P.M. George, B.K. Raghunath, L.M. Manocha, A.M. Warrier, EDM machining of carbon-carbon composite—a Taguchi approach. J. Mater. Process. Technol. **145**(1), 66–71 (2004)

41. H.K. Kansal, S. Singh, P. Kumar, Application of Taguchi method for optimisation of powder mixed electrical discharge machining. Int. J. Manuf. Technol. Manag. **7**(I), 329–341 (2005)

42. Y.C. Lin, C.H. Cheng, B.L. Su, L.R. Hwang, Machining characteristics and optimization of machining parameters of SKH 57 high-speed steel using electrical-discharge machining based on Taguchi method. Mater. Manuf. Process. **21**(8), 922–929 (2006)

43. H.K. Kansal, S. Singh, P. Kumar, Performance parameters optimization (multi-characteristics) of powder mixed electric discharge machining (PMEDM) through Taguchi's method and utility concept. Indian J. Eng. Mater. Sci. **13**(3), 209–216 (2006)

44. K.D. Chattopadhyay, S. Verma, P.S. Satsangi, P.C. Sharma, Development of empirical model for different process parameters during rotary electrical discharge machining of copper-steel (EN-8) system. J. Mater. Process. Technol. **209**(3), 1454–1465 (2009)

45. P. Govindan, S.S. Joshi, Experimental characterization of material removal in dry electrical discharge drilling. Int. J. Mach. Tools Manuf. **50**(5), 431–443 (2010)

46. S.R. Nipanikar, D.V. Ghewade, Electro discharge machining of Inconel material. Int. J. Eng. Res. Technol. **4**(2), 157–169 (2011)

47. K.H. Syed, K. Palaniyandi, Performance of electrical discharge machining using aluminium powder suspended distilled water. Turkish J. Eng. Env. Sci. **36**, 195–207 (2012)

48. D.R. Vhatkar, B.R. Jadhav, An experimental study on parametric optimization of high carbon steel (EN-31) by using silicon powder mixed dielectric EDM process. Int. J. Sci. Eng. Res. **1**(1–3), 431–436 (2013)

49. R.C. Amit, J. Garg, P.P. Sing, Experimental investigation of process parameters on EN-5 Steel using EDM drilling and Taguchi Method. Int. J. Sci. Eng. Technol. Res. **2**(11), 2049–2051 (2013)

50. A. Bergaley, N. Sharma, Optimization of electrical and non electrical factors in EDM for machining die steel Using Copper electrode by adopting Taguchi technique. Int. J. Innov. Technol. Explor. Eng. **3**(3), 44–48 (2013)

51. S. Goyal, R.K. Singh, Parametric study of powder mixed EDM and optimization of MRR & surface roughness. Int. J. Sci. Eng. Technol. **3**(1), 56–62 (2014)

52. S. Nagaraja, K. Chandrasekaran, S. Shenbhgaraj, Optimization of parameter for metal matrix composite in wire EDM. Int. J. Eng. Sci. Res. Technol. **4**(2), 570–574 (2015)

53. C.S. Kalra, L. Singh, S. Singh, Experimental investigation on surface roughness of heat treated Ti-6Al-4V machined by WEDM. Int. J. Manuf. Mech. Eng. **2**(December), 1–10 (2016)

54. A. Kumar, H. Majumder, K. Vivekananda, K.P. Maity, NSGA-II approach for multi-objective optimization of wire electrical discharge machining process parameter on Inconel 718. Mater. Today Proc. **4**(2), 2194–2202 (2017)

55. S.D. Mohanty, R.C. Mohanty, S.S. Mahapatra, Study on performance of EDM electrodes produced. J. Adv. Manuf. Syst. **16**(4), 357–374 (2017)

56. J.S. Soni, G. Chakraverti, Machining characteristics of titanium with rotary electro-discharge machining. Wear **171**(1–2), 51–58 (1994)

57. I. Puertas, C.J. Luis, L. Álvarez, Analysis of the influence of EDM parameters on surface quality, MRR and EW of WC-Co. J. Mater. Process. Technol. **153–154**(1–3), 1026–1032 (2004)

58. H.K. Kansal, S. Singh, P. Kumar, Parametric optimization of powder mixed electrical discharge machining by response surface methodology. J. Mater. Process. Technol. **169**(3), 427–436 (2005)

59. K.T. Chiang, F.P. Chang, D.C. Tsai, Modeling and analysis of the rapidly resolidified layer of SG cast iron in the EDM process through the response surface methodology. J. Mater. Process. Technol. **182**(1–3), 525–533 (2007)

60. C.J. Luis, I. Puertas, Methodology for developing technological tables used in EDM processes of conductive ceramics. J. Mater. Process. Technol. **189**(1–3), 301–309 (2007)

61. P. Kuppan, A. Rajadurai, S. Narayanan, Influence of EDM process parameters in deep hole drilling of Inconel 718. Int. J. Adv. Manuf. Technol. **38**(1–2), 74–84 (2008)

62. K.T. Chiang, Modeling and analysis of the effects of machining parameters on the performance characteristics in the EDM process of Al_2O_3 + TiC mixed ceramic. Int. J. Adv. Manuf. Technol. **37**(5–6), 523–533 (2008)

63. K M. Patel, P.M. Pandey, P. Venkateswara Rao, Determination of an optimum parametric combination using a surface roughness prediction model for EDM of Al 2 O 3/SiC w/TiC ceramic composite. Mater. Manuf. Process. **24**(6), 675–682 (2009)
64. M.S. Sohani, V.N. Gaitonde, B. Siddeswarappa, A.S. Deshpande, Investigations into the effect of tool shapes with size factor consideration in sink electrical discharge machining (EDM) process. Int. J. Adv. Manuf. Technol. **45**(11–12), 1131–1145 (2009)
65. S.K. Saha, S.K. Choudhury, Experimental investigation and empirical modeling of the dry electric discharge machining process. Int. J. Mach. Tools Manuf **49**(3–4), 297–308 (2009)
66. S.S. Habib, Study of the parameters in electrical discharge machining through response surface methodology approach. Appl. Math. Model. **33**(12), 4397–4407 (2009)
67. M.K. Pradhan, C.K. Biswas, Modeling and analysis of process parameters on Surface Roughness in EDM of AISI D2 tool steel by RSM approach," vol. 33, no. November, pp. 814–819 (2009)
68. A.K. Iqbal, A.A. Khan, Modeling and analysis of MRR, EWR and surface roughness in EDM milling through response surface methodology. Am. J. Eng. Appl. Sci. **3**(4), 611–619 (2010)
69. B. Jabbaripour, M.H. Sadeghi, S. Faridvand, M.R. Shabgard, Investigating the effects of EDM parameters on surface integrity, MRR and TWR in machining of Ti-6Al-4V. Mach. Sci. Technol. **16**(3), 419–444 (2012)
70. P. Shandilya, P.K. Jain, N.K. Jain, Prediction of surface roughness during wire electrical discharge machining of SiC p/6061 Al metal matrix composite. Int. J. Ind. Syst. Eng. **12**(3), 301–315 (2012)
71. S. Assarzadeh, M. Ghoreishi, Statistical modeling and optimization of process parameters in of cobalt-bonded tungsten carbide composite (WC/6%Co). Procedia CIRP **6**, 463–468 (2013)
72. S. Rajendran, K. Marimuthu, M. Sakthivel, Study of crack formation and resolidified layer in EDM process on T90Mn2W50Cr45 tool steel. Mater. Manuf. Process. **28**(6), 664–669 (2013)
73. I. Ayesta et al., Influence of EDM parameters on slot machining in C1023 aeronautical alloy. Procedia CIRP **6**, 129–134 (2013)
74. K.H. Syed, P. Kuppan, Studies on recast-layer in EDM using aluminium powder mixed distilled water dielectric fluid. Int. J. Eng. Technol. **5**(2), 1775–1780 (2013)
75. K. Gupta, N.K. Jain, Analysis and optimization of surface finish of wire electrical discharge machined miniature gears. J. Eng. Manuf. (2013)
76. G. Talla, S. Gangopadhyay, C.K. Biswas, Influence of graphite powder mixed EDM on the surface integrity characteristics of Inconel 625. Part. Sci. Technol. **6351**(March) (2016)
77. P. Shandilya, P.K. Jain, N.K. Jain, Modelling and process optimisation for wire electric discharge machining of metal matrix composites. Int. J. Mach. Mach. Mater. **18**(4), 377–391 (2016)
78. M. Bhaumik, K. Maity, Multi objective optimization of PMEDM using response surface methodology coupled with fuzzy based desirability function approach. Decis. Sci. Lett. **6**, 387–394 (2017)
79. N. Kuruvila, H.V. Ravindra, Parametric influence and optimization of wire-EDM on hot die steel. Mach. Sci. Technol. **0344**, 47–75 (2011)
80. S. Datta, S.S. Mahapatra, Modeling, simulation and parametric optimization of wire EDM process using response surface methodology coupled with grey-Taguchi technique. Int. J. Eng. Sci. Technol. **2**(5), 162–183 (2010)

Chapter 2
Optimization of Machining Parameters for Material Removal Rate and Machining Time While Cutting Inconel 600 with Tungsten Carbide Textured Tools

M. Adam Khan and Kapil Gupta

Abstract Machinability of nickel-based superalloy Inconel 600 was investigated with textured tungsten carbide cutting tool. The cutting tool edges are modified with laser engraving process to produce three different texture patterns on flank face of the tool. Experiments are planned to performed on CNC machine with twenty-seven combinations of input process parameters. Cutting speed (50, 100 and 150 m/min), feed rate (0.08, 0.1 and 0.12 mm/rev), depth of cut (0.1, 0.2 and 0.3 mm) and tool texture (dimple, line and spline) are the input parameter to study the machinability in terms of material removal rate and machining time. To identify the optimal process parameter, three different optimization techniques namely TOPSIS, Grey relational analysis and MOORA have been employed. It has been identified that the maximum cutting speed of 150 m/min at feed rate of 0.12 mm/rev and 0.3 mm depth cut can produce maximum material removal rate of 4268.39 mm^3/min. Moreover, the optimum machining time obtained is 0.259 min.

Keywords Machining · MRR · Optimization · Superalloy · TOPSIS · MOORA

2.1 Introduction

Nickel-based superalloys are in high demand for many engineering applications due to their superior properties such as metallurgical stability, high strength and temperature resistance, and high corrosion resistance etc. [1, 2]. On the other hand, these alloys are difficult-to-machine by conventional cutting techniques that results in poor machinability in terms of high tool wear, surface quality deterioration, and high consumption or energy and resources etc. [3–6]. Ample research work has been done on machining of superalloys with a wide range of cutting tools and using

M. Adam Khan · K. Gupta (✉)
Mechanical and Industrial Engineering Technology, University of Johannesburg, Johannesburg, South Africa
e-mail: kgupta@uj.ac.za

© Springer Nature Switzerland AG 2021
S. Pathak (ed.), *Intelligent Manufacturing*, Materials Forming, Machining and Tribology, https://doi.org/10.1007/978-3-030-50312-3_2

Table 2.1 Chemical composition of the work material Inconel 600

Elements	Ni	Cr	Fe	C	Mn	S	Si
wt%	72	17	9	0.15	1.0	0.35	0.5

various optimization techniques for machinability enhancement of various difficult-to-machine materials [7–12]. Not only plain tools, but textured tools also have been used to enhance the machinability in recent past. The commonly available tool texture patterns are dimples, holes, grooves and tracks in different angles on tool flank and rake faces [12–14]. There are different techniques followed to make tool texture on the cutting tools. In current research scenario, the tool texturing is made from laser engraving, ion beam texture process, electrochemical processing and wire-EDM techniques [12–18]. Laser surface texturing or engraving has high demand and remarkable benefits such as low cost, less time, high speed, eco-friendliness.

In the current research, an attempt has been made to enhance the machinability of Inconel 600 using laser engraved tools. The material removal rate (MRR) and cutting time are calculated using standard formulas corresponding to each experimental run. Further, three important and most extensively used optimisation techniques i.e. TOPSIS, Grey relational analysis, and MOORA have been adopted to find the optimal solutions for the best values of MRR and machining time.

2.2 Experimental Procedure

2.2.1 Materials

The work material Inconel 600 used in the present work is a grade of high performance nickel-chromium (Ni-Cr) based alloy with superior mechanical properties and resistance for high temperature and chemical environments. Nominal chemical composition of the work material is given in Table 2.1. The diameter of the rod used is 31 mm to a length of 3100 mm.

2.2.2 Cutting Tool and Texturing

Commercially available SANDVIK tools CNMG12-04-04 model, cadmium-nickel coated tungsten carbide material has been used to machine the Inconel 600 alloy. The hardness of the cutting tool material 1114Hv was four times higher than the work material. To modify the cutting tool surface area, solid state laser (Nd-YAG rod) machine (Make: Lee Laser Machine, USA; Model: SLT Q905) is used for engraving (texturing) on the cutting edge. Different texture patterns are made on the flank face by engraving to a depth of 50 μm for an area of 2.5 × 2.5 mm^2. Table 2.2 shows

Table 2.2 Details on cutting tool and texture specifications

Details	Specifications
Cutting tool	Cd–Ni coated WC insert
Tool model	CNMG12-04-04
Texture face	Flank face
Texture area	2.5 × 2.5 mm
Texture depth	50 μm
Texture pattern	Dimple, line and spline

Table 2.3 Laser process parameters used for texture patterning on cutting tool inserts

Parameter	Range	Units
Laser power	3800–4000	W
Laser travel speed	50	mm/s
Type of gas	Argon	–
Gas pressure	2.5	bar
Nozzle diameter	0.25	mm
Nozzle stand-off distance	25	mm
Focus position	Vertical	–

the detailed information of different texture patterns, whereas Table 2.3 presents the laser parameters used for texturing. Textures are made close to the cutting nose with a clearance of 150 μm on flank face. Table 2.4 presents the input machining process parameters.

Experiments have been performed on FANUC CNC machine (automatic lathe) with a cutting edge modified coated tungsten carbide inserts. The machining studies are performed by varying the cutting process parameters cutting speed, feed rate and depth of cut along with texture pattern. The aim of this research is to study the material removal rate for a straight cutting length of 50 mm (constant) at different process conditions.

Twenty seven experiments have been planned based on Taguchi L_{27} orthogonal array considering the input machining parameters such as; cutting speed, feed rate, depth of cut and texture pattern. Equation 2.1 has been used to evaluate the MRR

Table 2.4 Input process parameters and the range proposed for the research

Input parameters	Unit	Level 1	Level 2	Level 3
Cutting speed	m/min	50	100	150
Feed rate	mm/rev	0.08	0.1	0.12
Depth of cut	mm	0.1	0.2	0.3
Texture pattern	–	Dimple	Line	Spline

that is one of the machinability indicators. In addition to material removal rate the average machining time (Eq. 2.2) has also been calculated with the help of Eqs. (2.3) and (2.4). Following mathematical relations have been used for the calculations [19]:

$$\text{Material removal rate} = v_{avg}.fd \tag{2.1}$$

$$\text{Average Machining time } (T_m) = l/(fN) \tag{2.2}$$

$$\text{Average cutting speed } \left(v_{avg}\right) = \pi D_{avg} N \tag{2.3}$$

$$\text{Average Diameter } \left(D_{avg}\right) = \left(D_o - D_f\right)/2 \tag{2.4}$$

where; length of cut (l), original diameter (D_o), final diameter (D_f), cutting speed (N), feed rate (f) and depth of cut (d).

2.3 Modelling and Optimization

The empirical modelling of experimental results has also been done using regression analysis. Analysis of variance (ANOVA) has been done to identify the influence of input process parameters. From the analysis, contribution of individual process parameters is evaluated. Subsequently, the research has been proposed to find optimal process condition using different techniques. The TOPSIS, grey relational analysis (GRA) and MOORA are the three different techniques proposed to identify the optimal process parameter for the best machinability.

The TOPSIS is one of the recent techniques to find the best optimal process condition from the proposed experiment design. It is a 'Technique for Order Preference by Similarity to an Ideal Solution' (TOPSIS) introduced by Hwang during 1981 [20, 21]. It can produce best results for both best hypothetical (maximum) and worst (minimum) process conditions. Hint of this techniques is to locate closeness coefficient between the feasible and ideal solution. Research articles are available to explain the basic information and standard procedure in TOPSIS method [20, 21]. In addition to TOPSIS techniques, the MOORA and GRA are also considered for process optimization. Vishal (2014) investigated and explored research to study the material removal rate using mathematical modelling techniques [22]. Analytical modelling assists researchers to find the best optimal solution from the experimental design. In some other important research work, grey relational analysis was found very effective for multi-objective optimization [23, 24]. The uncertainties and difficulties found (on decision making) in multi response optimization problem are controlled with grey relational analysis [25]. In addition to conventional techniques, the multi-objective optimization on the basis of ratio analysis (MOORA) is recent and advanced technique followed for optimization purposes [26]. Brauers 2008 found that the MOORA is one of the best techniques for multi – objective decision-making optimisation problems [27]. The uncertainties intricated during the optimization are controlled with MOORA technique [28].

Procedure and steps involved in TOPSIS method

Step 1: Matrix normalization using standard formula given below

$$T_{ij} = \frac{x_{ij}}{\sqrt{\sum_{i=1}^{n} x_{ij}^2}}$$

T_{ij}—normalized value of experiment x_{ij}; (i.e. ith value of the jth experiment)
Step 2: Weightage for individual response
Based on the importance of the research proposed, the weightage can be assigned. It is also possible to find through some standard process. In this research, the weightage are given based on process contribution.
Step 3: Weighted normalized matrix calculation using following formula

$$V_{ij} = W_{ij} \times T_{ij}$$

where V_{ij} is the normalized value after considering the weightage W_{ij}.
Step 4: Ideal solution calculation from the weighted decision matrix. The ideal solutions are both positive (best) ideal solution (V^+) and negative (worst) solution (V^-) for each attributes/experiment.

$$V^- = \text{Mini./Maxi.} \left(V_1^-, V_2^-, V_3^-, V_4^- \ldots V_n^- \right)$$
$$V^+ = \text{Mini./Maxi.} \left(V_1^+, V_2^+, V_3^+, V_4^+ \ldots V_n^+ \right)$$

Step 5: Separation measured for every solution, as positive (best) ideal solution (S^+) and negative (worst) solution (S^-)

$$S_i^+ = \sqrt{\sum_{i=1}^{n} \left(V_{ij} - V_j^+ \right)^2}$$

$$S_i^- = \sqrt{\sum_{i=1}^{n} \left(V_{ij} - V_j^- \right)^2}$$

Step 6: Closeness coefficient value calculation using following equation

$$CCO = \frac{S_i^-}{S_i^+ + S_i^-}$$

Step 7: According to the closeness coefficient (CCO) value, the values are ranked in ascending order to find the optimal solution.

From the observations, the significant reasons are discussed with valid justifications on material removal and machining time. On the same, the influence of process parameters is correlated to the output and optimal solutions which are discussed in detail.

Procedure and steps in Grey relational analysis

Step 1: To calculate S/N ratio for the corresponding responses. Based on the expectation, the response may be larger—the—better and lower—the—better. S/N ratio (η) for each condition are calculated with following equations.

$$\text{S/N ratio } (\eta) = -10 \log_{10} \left(\frac{1}{n}\right) \sum_{i=1}^{n} \frac{1}{y_{ij}^2} \text{---for larger the best}$$

$$\text{S/N ratio } (\eta) = -10 \log_{10} \left(\frac{1}{n}\right) \sum_{i=1}^{n} y_{ij}^2 \text{---for smaller the best}$$

$$\text{S/N ratio } (\eta) = 10 \log_{10} \left(\frac{\mu^2}{\sigma^2}\right) \text{---for nominal the best}$$

where $\mu = \frac{y^1 + y^2 + y^3 + \cdots yn}{n}$, $\sigma^2 = \frac{\sum (y^i - \bar{y})^2}{n-1}$.
'n' is the number of replications and y_{ij} is the response observed over each experiment. Lager—the—best is sought for maximum material removal and the smaller the best is sought for less machining time.

Step 2: The response (y_{ij}) is normalizing to (Z_{ij}) between zero to one ($0 < Z_{ij} < 1$)

$$Z_{ij} = \frac{\max(y_{ij}, i = 1, 2, \ldots n) - y_{ij}}{\max(y_{ij}, i = 1, 2, \ldots n) - \min(y_{ij}, i = 1, 2, \ldots n)} \text{(for smaller the best)}$$

$$Z_{ij} = \frac{y_{ij} - \min(y_{ij}, i = 1, 2, \ldots n)}{\max(y_{ij}, i = 1, 2, \ldots n) - \min(y_{ij}, i = 1, 2, \ldots n)} \text{(for larger the best)}$$

It is highly recommended to use the S/N ratio value.

Step 3: To find the grey relational coefficient from the S/N ratio normalized values

$$\gamma(y_o(k), y_i(k)) = \frac{\Delta\min + \xi \Delta\max}{\Delta_{oj}(k) + \xi \Delta\max}$$

where, number of experiments (n) is represented from $j = 1, 2, \ldots$n and number of response (m) from $k = 1, 2, \ldots$m. The $y_o(k)$ reference sequence represented as ($y_o(k) = 1$, $k = 1, 2, \ldots$m) and specific comparison sequence is $y_j(k)$. Then the absolute value Δ_{oj} is the difference between $y_o(k)$, $y_i(k)$; $[\Delta_{oj} = \|y_o(k) - y_i(k)\|]$

$$\Delta \min = \min_{j \in i} \min_k \| y_o(k) - y_i(k) \| \text{ for smallest value}$$

$$\Delta \max = \max_{j \in i} \max_k \| y_o(k) - y_i(k) \| \text{ for larger value}$$

ξ is the distinguish coefficient defined between zero to one. This value may vary based on the practical needs and requirements.

Step 4: To generate the grey relational grade

$$\bar{\Upsilon} = \frac{1}{k} \sum_{i=1}^{m} \gamma_{ij}$$

The grey relational grade ($\bar{\Upsilon}$) for the jth experiment for the 'k' number of experiments.

Step 5: To find the optimal factor/process condition

Short the grey relational grade and rank in order. The higher relational grade referred to a best process/product designed and the condition as an optimal process condition.

Procedure and steps involved in MOORA technique

Step 1: To determine the objective with set of evaluation attributes from the design of experiments. In this condition, the cutting speed, feed rate and depth of cut are the attributes for the objective material removal rate and machining time optimization.

Step 2: To form decision matrix to represent response on input process parameters. The response measured/calculated for every combination of attributes (X_{ij}) is framed as given below:

$$X = \begin{bmatrix} x_{11} & x_{12} \cdots & x_{1n} \\ \vdots & \ddots & \vdots \\ x_{m1} & x_{m2} \cdots & x_{mn} \end{bmatrix}$$

where x_{ij} is the output measures of combinations j to an objective i, $i = 1, 2, \ldots n$ are combinations of objective and $j = 1, 2, \ldots m$ for inputs.

Step 3: To find the square root of sum of squares for each combination of attributes, following the given mathematical relation

$$X_{ij}^* = \frac{X_{ij}}{\sqrt{\sum_{j=1}^{m} X_{ij}^2}}$$

The X_{ij}^* will be a number in between the interval of zero to one ($0 < X_{ij}^* < 1$) represented as an normalized value.

Step 4: Assessment on normalized value
To find the optimal solution for multi objective condition, the normalized are (i) added for maximization problem and (ii) subtraction for minimization problem.

$$Y_i = \sum_{j=1}^{g} X_{ij}^* - \sum_{j=g+1}^{n} X_{ij}^*$$

where, g—No. of attributes to maximize; (n-g)—No. of attributes to minimize and Y_i—assessed normalized value for each combination of process conditions.
Step 5: Assessment on normalized value considered with weightage
It is important to relate normalized value with a defined weightage (W_j). The mathematical equation to relate the normalized value with weightage is given below:

$$Y_i = \sum_{j=1}^{g} W_j X_{ij}^* - \sum_{j=g+1}^{n} W_j X_{ij}^*$$

The weightage may be considered based on the importance of the attribute or can be calculated using entropy method.
Step 6: Ranking order for final decision
The predicted value of Y_i may be either positive or negative. The best combination of process parameter has maximum value Y_i and the worst condition will have a minimum value of Y_i.

2.4   Results and Discussion

2.4.1   Effect of Process Parameters

Figure 2.1 shows the machining chamber for cutting Inconel 600 alloy using texture tungsten carbide tools. The original and final diameter of the rod are measured at each and every experimental run, to calculate the material removal rate using standard mathematical equation (Eq. 2.1). The calculated material removal rate values (in mm³/min) for corresponding combination of process condition are presented in Table 2.5.

From the obtained results, the response for individual experiments are studied in detail. The calculated MRR value is varying between 412 and 4268 mm³/min. The maximum amount of material removal will be at maximum feed rate and depth of cut. The tool feed on cutting is reflecting the amount of distance in contact with work [29]. Dinesh et al. [30] claims that the depth of cut is also inducing the tool to remove maximum bulk material from the work. Since there is a progressive increase in MRR with respect to the increment in feed rate along with depth of cut at maximum cutting speed. To infer the importance of input parameters on the removal of bulk material,

Fig. 2.1 Picture of machining chamber

values are plot in graph as desired. Figure 2.2 shows the variation in removal of bulk material for a unit time in terms of minute. Graph was plotted to coordinates of depth of cut and MRR at different cutting velocities and feed rate. At a cutting speed of 50 m/min, for individual feed rate of 0.08, 0.1 and 0.12 mm/rev, the MRR values are plotted for varying depth of cut (0.1, 0.2 and 0.3 mm). It has been understood that the maximum removal of 1521.74 mm^3/min was found at a highest feed rate of 0.12 mm/rev. Subsequently the minimum removal of 412.72 mm^3/min was found at a lowest feed rate of 0.08 mm/rev. On other way at minimum depth of cut, the bulk removal was found with significant variations and wide variation at a maximum depth of cut.

As similar to 50 m/min with varying feed and depth of cut, the results observed at 100 m/min and 150 m/min revealed same increasing trend with respect to feed and depth of cut. At 100 m/min, the variations in bulk removal are in wide compared to other two cutting velocities and very close results at 150 m/min. All the above the maximum bulk removal was found at 150 m/min with a maximum feed rate of 0.12 mm/rev and 0.3 mm depth of cut. It is a common thumb rule in cutting mechanism which has been proven experimentally. At the highest process condition of cutting speed, feed rate and depth of cut; removal of bulk will be maximum.

To discuss in detail on cutting parameters and material removal rate, a contour graph was plotted (Fig. 2.3) to infer the interaction effect on MRR. While comparing the speed and feed rate, the material removal are normal and linear in changes. On comparing speed with depth of cut, the changes in MRR are slight contour curvature and found maximum at higher speed and depth of cut. At the same, with speed and feed rate, the changes found with less intensity. That is the speed and depth of cut can able to remove maximum materials on proposed machining parametric design. However, the influence and correlation between the input parameters can also be determined using empirical modelling. Using MINITAB software, the experimental results are empirically evaluated to identify the contribution of individual process parameters. The contributions of individual process parameter as predicted

Table 2.5 Calculated MRR value for individual trial runs

Exp No.	Cutting speed	Feed rate	Depth of cut	Tool texture	MRR
1	50	0.08	0.1	Dimple	412.72
2	50	0.08	0.2	Line	793.49
3	50	0.08	0.3	Spline	1174.26
4	50	0.1	0.1	Line	519.23
5	50	0.1	0.2	Spline	935.28
6	50	0.1	0.3	Dimple	1298.07
7	50	0.12	0.1	Spline	539.20
8	50	0.12	0.2	Dimple	998.52
9	50	0.12	0.3	Line	1521.74
10	100	0.08	0.1	Dimple	766.86
11	100	0.08	0.2	Line	1331.36
12	100	0.08	0.3	Spline	2076.92
13	100	0.1	0.1	Line	765.53
14	100	0.1	0.2	Spline	1930.47
15	100	0.1	0.3	Dimple	2496.3
16	100	0.12	0.1	Spline	1038.46
17	100	0.12	0.2	Dimple	2156.80
18	100	0.12	0.3	Line	3115.38
19	150	0.08	0.1	Dimple	1149.49
20	150	0.08	0.2	Line	2250.75
21	150	0.08	0.3	Spline	3376.12
22	150	0.1	0.1	Line	1205.76
23	150	0.1	0.2	Spline	2813.44
24	150	0.1	0.3	Dimple	3768
25	150	0.12	0.1	Spline	1688.06
26	150	0.12	0.2	Dimple	2652.67
27	150	0.12	0.3	Line	4268.39

by analysis of variance (ANOVA) are given in bar chart as shown in Fig. 2.4. The contribution percentages are as follows; cutting speed 43.35%, depth of cut 43.54%, feed rate 4.17% and the tool texture 0.01%. It proves that the level of contribution for cutting speed and depth of cut are significantly equal to produce maximum material removal. However, the influence of feed rate has found in single digit and there is no contribution for tool texturing. The evaluated results are under good to fit on line with R^2 value of 91.1% (error % 8.9%). It presents the statistical fitness of the data.

In this research, the performance of individual parameters is evaluated with reference to material removal rate. It is also an indication that how fast or slow the amount of material was removed from bar stock. Based on the machining time, the production

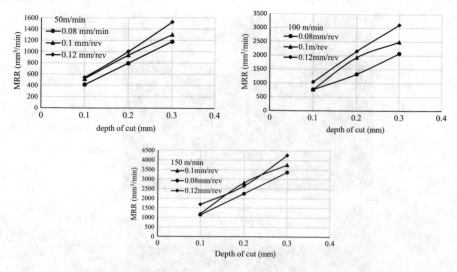

Fig. 2.2 Variation in material removal rate (MRR) with respect to depth of cut at different cutting velocities and feed rate

rate and its production cost are in demand. the machining time for a defined length of cut at different process conditions are mathematically calculated (Eq. 2.4). The analysis is made for a constant cutting length of 50 mm, nickel alloy (Inconel 600) rod. Figure 2.5 shows the average machining time calculated with reference to the diametral changes. The maximum time involved to cut the 50 mm length of superalloy is 1.2 min at machining condition of 0.08 mm/rev with 50 m/min cutting speed. As one the minimum cutting time in the same proposed set of experiment is 0.259 min while cutting the alloy at 0.12 mm/rev with 150 m/min cutting speed. When concluding the influence of process parameter for material removal and machining time, both are in contrast. Specifically, at a particular machining condition of 0.3 mm depth of cut, 0.12 mm/rev feed rate and cutting speed of 150 m/min; the maximum material removal is 4268.39 mm³/min at a minimum machining time of 0.259 min. The order of machining parameters may vary with reference to other cutting conditions. To rank the combination of cutting parameters and to find an optimal process condition, different optimisation techniques are proposed.

2.4.2 Results on Parameter Optimisation Using TOPSIS

Two responses i.e. material removal rate (MRR) and machining time (Tm) are considered and normalized. Weightage for these two responses are varied as 60% and 40% for MRR and Tm respectively. The significant reason for the range is that the MRR has variation in a wide range compared to Tm. Otherwise the direct result of input parameter (speed, feed and length of cut) are common for machining time. Therefore,

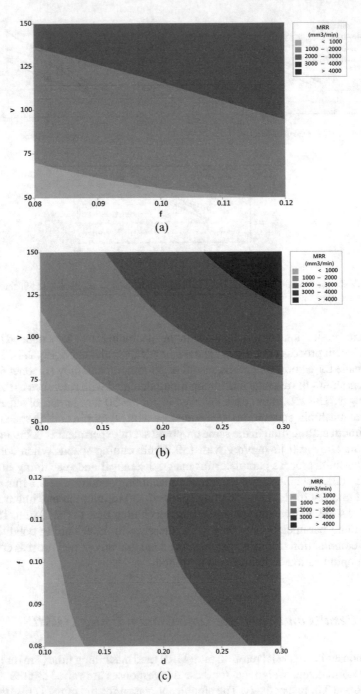

Fig. 2.3 Effects of parameter interactions **a** cutting speed v' and feed rate 'f', **b** cutting speed 'v' and depth of cut 'd' **c** feed rate 'f' and depth of cut 'd' on MRR

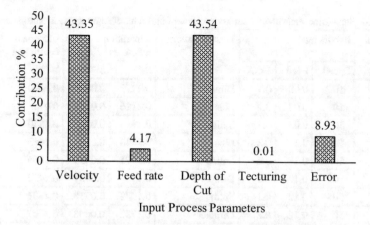

Fig. 2.4 Contribution of individual input process parameters towards material removal rate

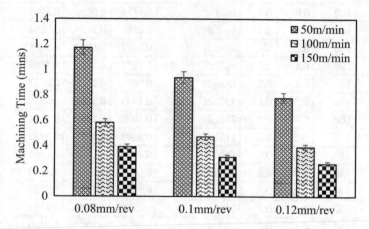

Fig. 2.5 Machining time calculated with reference to average diameter on machining and other input parameters

the weightage has been given more to MRR. Experimental responses are normalized and subsequently the separation measures and closeness coefficient were calculated and presented in Table 2.6. The optimal machining parameter obtained from the optimisation technique is 150 m/min, 0.12 mm/rev and 0.3 mm (v, f and d) in first order. Subsequently it shows that the cutting speed 150 m/min and 0.3 mm depth of cut can produce best response from the proposed result. Therefore, from the experimental investigations and process parameter optimisation, the machinability enhancement has been done to achieve the best MRR and it is at high speed condition.

Table 2.6 Separation measure, closeness coefficient and rank of input process parameters

Exp No.	Parameter			Texture pattern	Separation measures		Closeness coefficient	Rank
	Speed	Feed	DoC		Si^+	Si^-		
1	50	0.08	0.1	Dimple	0.2273	0.0000	0.0000	27
2	50	0.08	0.2	Line	0.2126	0.0181	0.0785	26
3	50	0.08	0.3	Spline	0.1988	0.0362	0.1540	24
4	50	0.1	0.1	Line	0.2043	0.0348	0.1455	25
5	50	0.1	0.2	Spline	0.1871	0.0429	0.1865	23
6	50	0.1	0.3	Dimple	0.1724	0.0551	0.2422	21
7	50	0.12	0.1	Spline	0.1932	0.0578	0.2301	22
8	50	0.12	0.2	Dimple	0.1732	0.0643	0.2707	20
9	50	0.12	0.3	Line	0.1511	0.0786	0.3423	18
10	100	0.08	0.1	Dimple	0.1733	0.0877	0.3361	19
11	100	0.08	0.2	Line	0.1476	0.0969	0.3963	16
12	100	0.08	0.3	Spline	0.1145	0.1174	0.5062	12
13	100	0.1	0.1	Line	0.1693	0.1047	0.3821	17
14	100	0.1	0.2	Spline	0.1153	0.1262	0.5225	11
15	100	0.1	0.3	Dimple	0.0896	0.1435	0.6157	7
16	100	0.12	0.1	Spline	0.1547	0.1186	0.4339	15
17	100	0.12	0.2	Dimple	0.1022	0.1418	0.5811	9
18	100	0.12	0.3	Line	0.0581	0.1726	0.7482	4
19	150	0.08	0.1	Dimple	0.1495	0.1204	0.4461	14
20	150	0.08	0.2	Line	0.0978	0.1447	0.5968	8
21	150	0.08	0.3	Spline	0.0464	0.1822	0.7969	3
22	150	0.1	0.1	Line	0.1457	0.1321	0.4754	13
23	150	0.1	0.2	Spline	0.0696	0.1705	0.7102	5
24	150	0.1	0.3	Dimple	0.0250	0.2038	0.8909	2
25	150	0.12	0.1	Spline	0.1226	0.1472	0.5456	10
26	150	0.12	0.2	Dimple	0.0768	0.1714	0.6907	6
27	**150**	**0.12**	**0.3**	**Line**	**0.0000**	**0.2273**	**1.0000**	**1**

2.4.3 Results on Parameter Optimisation Using Grey Relation Analysis

Using the standard grey relational analysis procedure, the response on material removal rate and machining time are analysed via a step-by-step process. The complete results arrived from the grey relational analysis is given in Table 2.7. The grey relational analysis for all the experimental trials, complete results on normalized value, grey co-efficient, grading and the ranking order are given. Similar to results

Table 2.7 Response and results arrived from grey relational analysis with rank order

Exp. No.	Process parameter(s)			Normalized value (N_{ij})		$1 - (N_{ij})$		Grey coefficient		Grade	Rank
	Speed	Feed	DoC	MRR	Tm	MRR	Tm	MRR	Tm		
1	50	0.08	0.1	0.0000	0.0000	1.0000	1.0000	0.5327	0.4288	0.4807	27
2	50	0.08	0.2	0.0988	0.0045	0.9012	0.9955	0.5702	0.4300	0.5001	26
3	50	0.08	0.3	0.1975	0.0090	0.8025	0.9910	0.6135	0.4313	0.5224	26
4	50	0.1	0.1	0.0276	0.2560	0.9724	0.7440	0.5427	0.5170	0.5298	25
5	50	0.1	0.2	0.1355	0.2600	0.8645	0.7400	0.5856	0.5187	0.5521	24
6	50	0.1	0.3	0.2296	0.2645	0.7704	0.7355	0.6290	0.5206	0.5748	22
7	50	0.12	0.1	0.0328	0.4271	0.9672	0.5729	0.5446	0.5994	0.5720	22
8	50	0.12	0.2	0.1519	0.4307	0.8481	0.5693	0.5927	0.6015	0.5971	21
9	50	0.12	0.3	0.2876	0.4340	0.7124	0.5660	0.6591	0.6033	0.6312	20
10	100	0.08	0.1	0.0918	0.6401	0.9082	0.3599	0.5674	0.7479	0.6577	19
11	100	0.08	0.2	0.2383	0.6430	0.7617	0.3570	0.6333	0.7505	0.6919	18
12	100	0.08	0.3	0.4316	0.6453	0.5684	0.3547	0.7479	0.7525	0.7502	16
13	100	0.1	0.1	0.0915	0.7686	0.9085	0.2314	0.5673	0.8793	0.7233	16
14	100	0.1	0.2	0.3936	0.7699	0.6064	0.2301	0.7222	0.8808	0.8015	14
15	100	0.1	0.3	0.5404	0.7725	0.4596	0.2275	0.8327	0.8840	0.8583	12
16	100	0.12	0.1	0.1623	0.8536	0.8377	0.1464	0.5973	0.9949	0.7961	13
17	100	0.12	0.2	0.4523	0.8551	0.5477	0.1449	0.7627	0.9973	0.8800	11
18	100	0.12	0.3	0.7010	0.8569	0.2990	0.1431	1.0000	1.0000	1.0000	7
19	150	0.08	0.1	0.1911	0.8561	0.8089	0.1439	0.6105	0.9988	0.8046	10
20	150	0.08	0.2	0.4767	0.8577	0.5233	0.1423	0.7809	1.0012	0.8910	8
21	150	0.08	0.3	0.7686	0.8592	0.2314	0.1408	1.0925	1.0036	1.0480	5
22	150	0.1	0.1	0.2057	0.9411	0.7943	0.0589	0.6173	1.1507	0.8840	7
23	150	0.1	0.2	0.6226	0.9421	0.3774	0.0579	0.9107	1.1528	1.0318	5
24	150	0.1	0.3	0.8702	0.9438	0.1298	0.0562	1.2688	1.1562	1.2125	2
25	150	0.12	0.1	0.3308	0.9974	0.6692	0.0026	0.6834	1.2796	0.9815	4
26	150	0.12	0.2	0.5809	0.9990	0.4191	0.0010	0.8694	1.2836	1.0765	3
27	**150**	**0.12**	**0.3**	**1.0000**	**1.0000**	**0.0000**	**0.0000**	**1.5981**	**1.2863**	**1.4422**	**1**

from the TOPSIS algorithm, the trial experiment number 27 found with the optimised process condition. The best results from the experimental design are found with the plan of high-speed machining process (150 m/min). Therefore, it is confirmed that the 150 m/min–0.12 mm/rev–0.3 mm can produce best machining response on machining hard to cut nickel-based superalloy.

2.4.4   Results on Parameter Optimisation Using MOORA

Table 2.8 shows the process parameter rank achieved in step-by-step procedure from multi-objective optimization on the basis of ratio analysis (MOORA) technique. The weightage for MRR and machining time is maintained same (60% for MRR and 40% for Tm) as in TOPSIS algorithm. In MOORA too, the optimal machining condition has been obtained corresponding to Experiment number 27 as seen in Table 2.8.

2.4.5   Discussion on Optimal Process Parameter

From the three optimization techniques used in the present work, the optimal process conditions in the first three rank order are given in Table 2.9. In all techniques, the optimal solution for the highest material removal rate is the combination of parameters 150 m/min—0.12 mm/rev—0.3 mm. Ramu et al. [31] confirms that, while using the multi response optimization techniques the grey relational analysis can produce maximum material removal at higher process condition. These techniques do not have any complication while optimising using GRA analysis [32]. While machining any material at a high speed, the plastic deformation of the material is high and when the depth of cut is increased the rate of deformation will be rapid. Significantly, the maximum feed rate has highly influenced the bulk deformation of material and machining time.

2.5   Conclusions

In this chapter, machinability investigation of Inconel 600 superalloy is reported. MRR and machining time are the two machinability indicators considered. TOPSIS, Grey, and MOORA based optimization has been done. Following conclusions can be drawn from this work:

1. From the analysis it is found that the variation in material removal rate completely depends on the feed rate followed by depth of cut and the cutting speed. Experiments confirmed that, the maximum travel of tool can cut extreme amount of bulk from the bar stock. At that moment, the increase in depth of cut is highly supportive to rise the material removal. From the design of experiments, factorial analysis confirms that the depth of cut is the predominant factor (43.54%), highly influencing the removal of bulk from the bar stock.
2. The variation in machining time for a particular cutting speed with different feed rate and depth of cutting is in the range of $\pm 5\%$. The minimum time required for defined cutting length of 50 mm is 0.259 min, at a maximum cutting speed of 150 m/min and the corresponding material removal is maximum ($4268.39 \text{ mm}^3/\text{min}$).

Table 2.8 Normalized decision matrix and rank order for MOORA

Exp No.	Process parameter(s)			Square of response X_i		Normalized value of X_i		Weightage $X_i * j$		$\Sigma Max - \Sigma min$	Rank
	v	f	d	MRR	Tm	MRR	Tm	MRR	Tm		
1	50	0.08	0.1	170,339.12	1.381659	0.039	0.344	0.024	0.138	0.161	17
2	50	0.08	0.2	629,627.27	1.372014	0.075	0.343	0.045	0.137	0.182	10
3	50	0.08	0.3	1,378,885.42	1.362379	0.112	0.342	0.067	0.137	0.204	5
4	50	0.1	0.1	269,600.21	0.884296	0.049	0.275	0.030	0.110	0.140	20
5	50	0.1	0.2	874,749.43	0.877369	0.089	0.274	0.053	0.110	0.163	16
6	50	0.1	0.3	1,685,001.30	0.869575	0.123	0.273	0.074	0.109	0.183	9
7	50	0.12	0.1	290,737.50	0.613481	0.051	0.229	0.031	0.092	0.123	22
8	50	0.12	0.2	997,042.19	0.608197	0.095	0.228	0.057	0.091	0.148	18
9	50	0.12	0.3	2,315,706.26	0.603543	0.145	0.228	0.087	0.091	0.113	23
10	100	0.08	0.1	588,079.41	0.34525	0.073	0.172	0.044	0.069	0.178	12
11	100	0.08	0.2	1,772,519.45	0.34212	0.127	0.171	0.076	0.069	0.144	19
12	100	0.08	0.3	4,313,603.33	0.339679	0.197	0.171	0.118	0.068	0.187	8
13	100	0.1	0.1	586,039.24	0.220571	0.073	0.138	0.044	0.055	0.099	27
14	100	0.1	0.2	3,726,722.14	0.219445	0.183	0.137	0.110	0.055	0.165	15
15	100	0.1	0.3	6,231,513.69	0.217193	0.237	0.137	0.142	0.055	0.197	6
16	100	0.12	0.1	1,078,400.83	0.153327	0.099	0.115	0.059	0.046	0.105	26
17	100	0.12	0.2	4,651,800.04	0.152233	0.205	0.114	0.123	0.046	0.169	14
18	100	0.12	0.3	9,705,607.50	0.150971	0.296	0.114	0.178	0.046	0.223	4
19	150	0.08	0.1	1,321,330.02	0.151523	0.109	0.114	0.066	0.046	0.111	24
20	150	0.08	0.2	5,065,884.57	0.150412	0.214	0.114	0.128	0.045	0.174	13

(continued)

Table 2.8 (continued)

Exp No.	Process parameter(s)			Square of response X_i		Normalized value of X_i		Weightage $X_i *_j$		ΣMax $-$ Σmin	Rank
	V	f	d	MRR	Tm	MRR	Tm	MRR	Tm		
21	150	0.08	0.3	11,398,240.27	0.149336	0.321	0.113	0.193	0.045	0.238	3
22	150	0.1	0.1	1,453,857.18	0.096845	0.115	0.091	0.069	0.036	0.105	25
23	150	0.1	0.2	7,915,444.63	0.096267	0.267	0.091	0.160	0.036	0.197	7
24	150	0.1	0.3	14,197,824.00	0.095327	0.358	0.090	0.215	0.036	0.251	2
25	150	0.12	0.1	2,849,560.07	0.067335	0.160	0.076	0.096	0.030	0.127	21
26	150	0.12	0.2	7,036,668.74	0.06659	0.252	0.076	0.151	0.030	0.181	11
27	**150**	**0.12**	**0.3**	**18,219,156.61**	**0.066106**	**0.406**	**0.075**	**0.243**	**0.030**	**0.274**	**1**

Table 2.9 Summary on optimal process parameters obtained by TOPSIS, Grey, and MOORA

Ranking order	TOPSIS	Grey relational analysis	MOORA
1st	150–0.12–0.3	150–0.12–0.3	150–0.12–0.3
2nd	150–0.1–0.3	150–0.1–0.3	150–0.1–0.3
3rd	150–0.08–0.3	150–0.12–0.2	150–0.08–0.3

3. From the empirical modelling, it is found that the optimal machining condition for the defined experimental plan is 150 m/min—0.12 mm/rev—0.3 mm. The TOPSIS, Grey relational analysis and MOORA techniques have confirmed the same high-speed machining condition as an ideal process condition to machine Inconel 600 type superalloy.

4. In essence, it can be suggested that while machining the hard nickel-based superalloy at higher cutting speed, feed rate and depth of cut, maximum amount of material can be removed with minimum machining time.

References

1. R. Joseph, *Davis Alloying: Understanding the Basics* (ASM International, Materials Park, Ohio, 2001)
2. A.P. Mouritz, *Introduction to Aerospace Materials* (Woodhead Publishing Limited, Cambridge, UK, 2012)
3. E.O. Ezugwu, Z.M. Wang, A.R. Machado, The machinability of nickel-based alloys: a review. J. Mater. Process. Technol. **86**, 1–16 (1999)
4. Y.B. Guo, W. Li, I.S. Jawahir, Surface integrity characterization and prediction in machining of hardened and difficult-to-machine alloys; a state-of-the-art research review and analysis. Mach. Sci. Technol. **13**, 437–70 (2009)
5. D.G. Thakur, B. Ramamoorthy, L. Vijayaraghavan, Study on the machinability characteristics of superalloy Inconel 718 during highspeed turning. Mater. Des. **30**, 1718–1725 (2009)
6. Salman Pervaiz, Amir Rashid, Ibrahim Deiab, Mihai Nicolescu, Influence of tool materials on machinability of titanium- and nickel-based alloys: a review. Mater. Manuf. Processes **29**, 219–252 (2014)
7. A. Sharman, R.C. Dewes, D.K. Aspinwall, Tool life when high speed ball nose end milling Inconel. J. Mater. Process. Technol. **118**, 29–35 (2001)
8. S.A. Khan, S.L. Soo, D.K. Aspinwall, C. Sage, P. Harden, M. Fleming, A. White, R. M'Saoubi, Tool wear/life evaluation when finish turning Inconel 718 using PCBN tooling. Procedia CIRP **1**, 283–288 (2012)
9. V. Bushlya, J. Zhou, J.E. Ståhl, Effect of cutting conditions on machinability of superalloy Inconel 718 during high speed turning with coated and uncoated PCBN tools. Procedia CIRP **3**, 370–375 (2012)
10. T. Ozel, E. Zeren, A methodology to determine work material flow stress and tool-chip interfacial friction properties by using analysis of machining. J. Manuf. Sci. Eng. **128**, 119–129 (2006)
11. G.C. Manjunath Patel, D. Lokare, G. Chate, M.B. Parappagoudar, R. Nikhil, K. Gupta, Analysis and optimization of surface quality while machining high strength aluminium alloy. Measurement (Elsevier), 107337 (2020). https://doi.org/10.1016/j.measurement.2019.107337
12. P.M. Mashinini, H. Soni, K. Gupta, Investigation on dry machining of stainless steel 316 using textured tungsten carbide tools. Mater. Res. Express (IOP Science) **7**, 016502 (2020). https://doi.org/10.1088/2053-1591/ab5630

13. N. Kawasegi, H. Sugimori, N. Morita, M. Xue, Atmosphere effects on the machinability of cutting tools with micro- and nanoscales textures. Adv. Mater. Res. **325**, 333–338 (2011)
14. T. Sugihara, T. Enomoto, Crater and flank wear resistance of cutting tools having micro textured surfaces. Precis. Eng. **37**, 888–896 (2013)
15. Z. Wu, J.Z. Deng, Y.Q. Xing, H. Cheng, J. Zhao, Effect of surface texturing on friction properties of WC/Co cemented carbide. Mater. Des. **41**, 142–149 (2012)
16. I. Etsion, State of the art in laser surface texturing. J. Trib. **127**(1), 761–762 (2005)
17. X.L. Chen, N.S. Qu, H.S. Li, Z.Y. Xu, Electrochemical micromachining of micro-dimple arrays using a polydimethylsiloxane (PDMS) mask. J. Mater. Process. Technol. **229**, 102–110 (2016)
18. K. Gupta, N. K. Jain, Deviations in geometry of miniature gears fabricated by wire electrical discharge machining. *Proceedings of the ASME 2013 International Mechanical Engineering Congress and Exposition*. vol.10, Micro-and Nano-Systems Engineering and Packaging. San Diego, California, USA. 15–21 Nov 2013. V010T11A047. ASME. https://doi.org/10.1115/IMECE2013-66560
19. M. Senthil Kumar, R. Adam Khan, T.Sornakumar Thiraviam, Machining parameters optimization for alumina based ceramic cutting tools using genetic algorithm. Mach. Sci. Technol. Int. J. **10**(4), 471–489 (2006)
20. S.S. Naik, J. Rana, P. Nanda, Using TOPSIS method to optimize the process parameters of D2 steel on electro-discharge machining. Int. J. Mech. Eng. Technol. **9**(13), 1083–1090 (2018)
21. A.K. Parida, B.C. Routara, *Multiresponse Optimization of Process Parameters in Turning of GFRP Using TOPSIS Method*, International Scholarly Research Notice, Hindawi Publisheing Co. (2014)
22. V. Francis, R.S. Singh, N. Singh, A.R. Rizvi, S. Kumar, Application of Taguchi method and ANOVA in optimization of cutting parameters for material removal rate and surface roughness in turning operation. Int. J. Mech. Eng. Technol. **4**, 47–53 (2014)
23. A.K. Sahoo, N.B. Achyuta, K.R. Arun, B.C. Routra, Multi-objective optimization and predictive modelling of surface roughness and material removal rate in turning using grey relational and regression analysis. Procedia Eng. **38**, 1606–1627 (2012)
24. S.H. Chang, J.R. Hwang, J.L. Doong, Optimization of the injection moulding process of short glass fiber reinforced polycarbonate composites using grey relational analysis. J. Mater. Process. Technol. **97**, 186–193 (2000)
25. S. Balasubramanian, S. Ganapathy, Grey relational analysis to determine optimum process parameters for wire electro discharge machining (WEDM). Int. J. Eng. Sci. Technol. **3**(1), 95–101 (2011)
26. W.K. Brauers, E.K. Zavadskas, Robustness of the multi-objective MOORA method with a test for the facilities sector. Technol. Econ. Dev. Econ. **15**(2), 352–375 (2009)
27. W.K.M. Brauers, Multiobjective contractor's ranking by applying the MOORA method. J. Bus. Econ. Manag. **4**, 245–255 (2008)
28. A. Muniappan, M. Sriram, C. Thiagarajan, G. Bharathi Raja, T. Shaafi, Optimization of WEDM process parameters on Machining of AZ91 magnesium alloy using MOORA method. IOP Conf. Ser. Mater. Sci. Eng. **390**(2018), 012107. https://doi.org/10.1088/1757-899x/390/1/012107
29. M.S. Ranganath, Vipin, R.S. Mishra, Effect of cutting parameters on MRR and surface roughness in turning of aluminium (6061). Int. J. Adv. Res. Innov. **2**(1), 32–39 (2014)
30. S. Dinesh, K. Rajaguru, V. Vijayan, A. Godwin Antony, Investigation and prediction of material removal rate and surface roughness in CNC turning of EN24 alloy steel. Mech. Mech. Eng. **20**(4), 451–466
31. I. Ramu, P. Srinivas, K. Vekatesh, Taguchi based grey relational analysis for optimization of machining parameters of CNC turning steel 316. IOP Conf. Ser.: Mater. Sci. Eng. **377**, 012078 (2018)
32. A. Noorul Haq, P. Marimuthu, R. Jeyapaul, Multi response optimization of machining parameters of drilling Al/SiC metal matrix composite using grey relational analysis in the Taguchi method. Int. J. Adv. Manuf. Technol. **37**, 250–255 (2008)

Chapter 3
Kurtosis Analysis of Tool Drilling Geometries and Cutting Conditions for Deep Twist Drilling Process Improvement

M. H. S. Harun and A. R. Yusoff

Abstract The chapter evaluates the tri-axial acceleration of different tool geometries and deep drilling cutting parameters using kurtosis analysis for improving the process. The solution is based on vibrations signal in tri-axial of x, y and z axes analysed by kurtosis analysis and the main effect and the analysis of variance (ANOVA) is used to improve the deep drilling process. The approach utilised experimental data from different tool geometries of point angle, helix angle and clearance angle and cutting parameters of feeds and cutting speed in drilling the die material of SKD 61. The kurtosis analysis was verified with process conditions of frequency analysis and time-frequency analysis. The result showed that the most significant parameters that affects the kurtosis acceleration are tool geometries of clearance angle, helix angle and point angle using ANOVA and main effect analyses. The optimal tool geometries are sharp point angle 115°, widen helix angle 35° and balanced clearance angle 120° from minimum kurtosis acceleration analysis and verified with signal processing analyses. The optimum tool geometries with process parameters can be determined based on minimum kurtosis value in drilling process can improve 93% compared to the worst tool geometry. It is concluded that kurtosis analysis can be used with tri-axial signal processing assistance to improve the tool geometry and cutting parameters performance in the deep drilling process.

Keywords Deep drilling process · Tool geometry · Main effect · Signal processing · Kurtosis analysis

M. H. S. Harun · A. R. Yusoff (✉)
Faculty of Manufacturing Engineering, Universiti Malaysia Pahang, 26600 Pekan, Pahang, Malaysia
e-mail: razlan@ump.edu.my

© Springer Nature Switzerland AG 2021
S. Pathak (ed.), *Intelligent Manufacturing*, Materials Forming, Machining and Tribology, https://doi.org/10.1007/978-3-030-50312-3_3

3.1 Introduction

Drilling is a type of machining process for making holes in solid materials. A survey in 1999 revealed that approximately 250 million twist drills are used annually by the US industry alone [1]. According to the US Department of Commerce [2], approximately $1.62 billion was spent in the production of drill bits in the United States in 1991. Thus, many drill bits are used, and much money is spent on advanced technology in the drilling process. Recently, deep drilling is used in die, engines, and aerospace industries [3]. Any hole ten times larger than its diameter is considered to be a deep hole. As a standard drill process, deep drilling also becomes one of the essential machining processes. A survey for 145 company in machining sector, they come out with results that showed the drilling process is one of the most important by 29% of machining processes distribution and 13% of machine time consumed [2]. At the same time, they also show evidence that deep drilling received high potential, due to the 7% of the deep drilling tasks mean around 28% of the processing time.

In tool drills mechanism there are several primary tool geometry that influences the cutting performance such as drill point/cutting lip angle, helix angle, lip clearance angle and chisel edge angle, as shown in Fig. 3.1. Each of the tool geometry listed had their contribution in tool drills cutting. Drill point angle is the angle between the drill lips and it directly related to the length of cutting edges, where the lowest point angle, means the highest length of cutting edges which make tool drill easily entered to the pilot hole. However, this technique is not free from limitations. One of the primary issues in deep twist drilling technique is premature tool breakage which is due to the tool wear, chip clogging and deflection, which can affect the quality of the process [1, 4]. This tool's failures mechanism can affect the productivity of the process. The demand on the optimum mechanism to undergo in-depth drilling process to produce high-quality product has forced researchers to overcome these issues.

Uysal et al. [4] found that the size of the point angle affects the condition of the tool as an example, 80° drill point angle worn less than 100° and 120° drill point angle. Next, the helix angle is the angle between the land's edge and the axis of the drill. In common usage, there are two types of well-known helix angle drills in industries which is high helix drills and low helix drills. High helix drill has a high helix angle which usually use to cut softer metals or low strength materials. Through its wide helix angle, it improves cutting efficiency but weakens the drill body. Meanwhile, low helix drill has a lower helix angle than a standard drill. It usually used to prevent the tool from running head or grabbing when drilling brass or similar materials. However based on author experienced, for deep drilling application, the high helix angle is more suitable. Wide flute allows chips to flow out smoothly out of the hole which can avoid chips clogged. But, it requires hard material such as solid carbide or high-speed steel as a drill body in order to avoid weakens the body.

Lip clearance angle is another parameter that influences the tool drill geometry, which it is high or low angles is affected by the cutting process flow. Settings of the lip clearance angle usually depend on hardness and machinability of the product

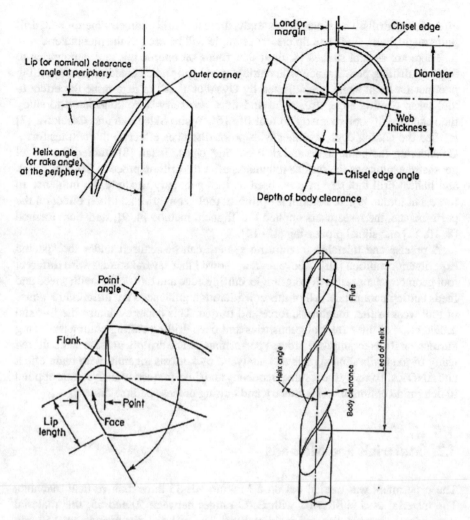

Fig. 3.1 Drill tool geometry [4]

material. Moreover, the chisel edge angle is the angle between the chisel edge and cutting lips where can be seen from the end of the drill. According to Mazoff [5], the higher number of chisel edge angle make chisel's profile flat (linear) which running straight across the drill, which make the chisel's full-length contacts metal and the chisel's corners simultaneously augurs into the metal and producing chips. Meanwhile, a low number of chisel angle produced a pronounced sharp edge on chisel which make chisel has outstanding extruding properties permitting less thrust and less heat. However, the chisel edge angle worked almost similar with the drill point angle, which by considering only one of them will be enough to cover their geometry effect in tool drills. Therefore, in this study, three tool drill geometry mentioned; drill point angle, helix angle and lip clearance angle will be used as the parameters.

There are several studies [4, 6–9] that focus on optimizing tool geometries to improve drilling performance with cutting speed and feeds consideration as cutting parameters. Point angle was adjusted by Uysal et al. [4] to determine the effect to tool wear. Cutting edge, drill point and flute geometry were considered to study the influence of tool geometry to tool life [6]. While Sklenicka and Cesakava [7] studied the cutting edge, together with point distances effect to the cutting force. Considering the point angle and drill coating types, Tosun [8] studied those tool geometries influence the surface roughness after the drilling process. Conical, racon and helical drill tool design were used as tool geometry to study the influence to force and torque [9]. To determine optimum tool geometries and their effect on the performance, the researchers applied the Taguchi method [4, 9], statistical method [8, 10, 11] and signal processing [12–14].

A precise and tolerable monitoring system can be realised under the specific experiment condition [10]. Above review showed that several studies with different tool geometry parameters of point angle, cutting edge and tool design with speed and feeds cutting parameters. All of this consideration influenced the different response of tool wear or life, roughness, force and torque. This paper evaluates the tri-axial acceleration of different tool geometries and deep drilling cutting parameters using kurtosis analysis for improve process performance. The drilling process is monitored using tri-axial vibration signals and analysed by kurtosis together with main effect and ANOVA. To verify the results, frequency and time-frequency analyses are applied to determine optimum tool geometry and cutting process parameters.

3.2 Materials and Methods

The experiment was carried out on a Makino KE-55 three-axis vertical machine. The material used is SKD 61 with HRC ranges between 50 and 55; this material is practically used in dies and mold making. The material dimensions are 150 mm × 150 mm × 80 mm, and the tool drill used is solid carbide tool drills material with 8 mm diameter and 120 mm length. By using the design of experiment (DOE), 27 experiment parameters were generated based on five factors with three levels of quarter factorial of DOE. Each of the holes is performed by using different tool geometry and cutting conditions, as listed in Table 3.1.

In terms of the data record, vibration sensors are mounted to the spindle head of the machine, as shown in Fig. 3.2. The tri-axial accelerometer (Model 356B21) with a sensitivity of ±10 mV/g is used. Recorded data are analysed by using a statistical parameter called kurtosis. Kurtosis was an established method and appropriate statistical parameter to represent low twist drill tool conditions for vibrations [10]. Based on Mahfouz [11], it is found that kurtosis was sensitive to the occurrence of spikes or impulses in the time domain of the vibration signal. Kurtosis, $K_{x,y,z}$ in the current experiment measured for x, y and z-axis directions can be defined as:

Table 3.1 Tool drills geometries and cutting conditions parameter for 27 experiment

No.	Point angle (°)	Helix angle (°)	Lip clearance angle (°)	Feeds (mm/rev)	Cutting speed (m/min)
1	115	30	110	0.1	30
2	115	30	110	0.1	40
3	115	30	110	0.1	50
4	115	35	120	0.2	30
5	115	35	120	0.2	40
6	115	35	120	0.2	50
7	115	40	130	0.3	30
8	115	40	130	0.3	40
9	115	40	130	0.3	50
10	120	30	120	0.3	30
11	120	30	120	0.3	40
12	120	30	120	0.3	50
13	120	35	130	0.1	30
14	120	35	130	0.1	40
15	120	35	130	0.1	50
16	120	40	110	0.2	30
17	120	40	110	0.2	40
18	120	40	110	0.2	50
19	125	30	130	0.2	30
20	125	30	130	0.2	40
21	125	30	130	0.2	50
22	125	35	110	0.3	30
23	125	35	110	0.3	40
24	125	35	110	0.3	50
25	125	40	120	0.1	30
26	125	40	120	0.1	40
27	125	40	120	0.1	50

$$K_{x,y,z} = \frac{1}{N} \sum_{i=1}^{N} \frac{(x_i - x)^4}{\sigma^4} \tag{3.1}$$

where x_i is average of data acceleration and x is average acceleration value, σ is standard deviation. Average K can be calculated based on K in x, y and z axis direction using Eq. (3.2)

$$K = \sqrt{K_x^2 + K_y^2 + K_z^2} \tag{3.2}$$

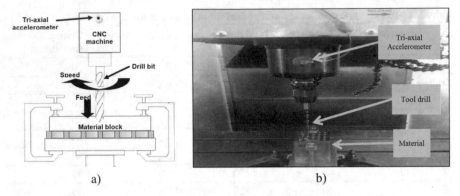

Fig. 3.2 The experimental setting for drilling **a** schematic diagram, **b** sensor arrangement

Optimum tool geometries parameters are determined from minimum kurtosis acceleration value using the main effect and analysis of variance (ANOVA).

In addition to time-domain analysis with kurtosis analysis, the measured signal affected due to tool wear phenomenon can be known through the implementation of frequency analysis and time-frequency analysis. Fast Fourier Transform (FFT) computes from the discrete Fourier transform of a sequence or its inverse. It is defined as follows:

$$F\{s(t)\}(\omega) = \int\limits_{-\infty}^{\infty} e^{(-i\omega t)}s(\tau)d\tau \qquad (3.3)$$

where symbol 's' in Eq. (3.3) represents the function of the Fourier transform. Meanwhile 'ω' is referred to single-frequency parameters that determine the basis function in the family and τ is referred to the time axis of the signal.

Frequency domain analyses the signal in term of frequency without knowing time happened during that frequency. To know time and frequency, short-time Fourier transform (STFT) is used to monitor and analyse the signal. STFT is based on FFT as a time-frequency technique to deal with non-stationary signals that have a short data window centred on time [15–17]. It is added a sliding window function $w(t)$ to FFT and obtains a localized time-frequency atom and the combination of the transform become STFT. The window base functions are defined as:

$$b_{(w,t_0)}(t) = w(t - t_0)e^{i\omega t} \qquad (3.4)$$

where $w(t)$ as mention before is the window functions, and (w, t_0) is the time-frequency coordinates of the base function. By adding a time dimension to the base functions parameters, the multiplying the infinitely long complex exponential with a window to localize it. Base window function, Eq. (3.4) to be multiplied to FFT Eq. (3.5) as described as:

$$\mathcal{S}\{s(t)\}(w, t_0) = \int\limits_{-\infty}^{\infty} w(\tau)e^{(-i\omega t)s(\tau)d\tau}$$ (3.5)

where $s(t)$ represented as a Fourier transform signal.

3.3 Results and Discussion

In this section, the results of tool performance are discussed in detail. Section 3.1 presents a statistical analysis of kurtosis. Section 3.2 describes the analysis of FFT and STFT explicitly. The data analysis of the optimisation of tool geometry is presented through the main effect analysis and analysis of variance (ANOVA) and Sect. 3.4 tackles the optimum tool selection before the discussion is made.

3.4 Kurtosis Analysis

The time-domain data for each axis, namely, x, y, and z, as sampling are analysed through the statistical method of kurtosis. Figure 3.3 shows a summary of the summation of kurtosis values for the 27 experiments in a bar chart. The sum

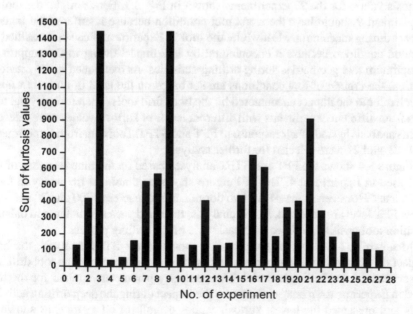

Fig. 3.3 Summary of kurtosis values for the 27 experiments

of the kurtosis values for each experiment is added from the kurtosis value in each drill axis, namely, x, y, and z. These three axes are proven to directly and effectively contribute to showing the effect of cutting parameter during the in-depth drilling process, especially in the y and z axes [10].

The highest number of kurtosis recorded was from Experiment 9 with a value 1451, followed by Experiment 3 (1450) and then Experiment 21 (915). Meanwhile, the lowest number of kurtosis recorded was from Experiment 4 with a value of 39, followed by Experiment 5 (56) and then by Experiment 10 (74). In a drilling process, the number of kurtosis represents the impact that tool drills encountered. The higher the impact encountered by drills during the drilling process, the bigger the vibration that occurs and the higher the kurtosis number. However, Fig. 3.3 shows that the number of kurtosis without considering the effect of each parameter in shaping drill performance lead to improper decision and condition. Therefore, the author discusses in detail the performance of each tool drill according to their parameters. To systematize the discussion, three regions were divided based on Experiments 1–9, 10–18, and 19–27, which are referred to as regions 1, 2, and 3, respectively.

3.4.1 Analysis of Tool Using Frequency Domain and Time-Frequency

Several criteria of tool conditions can be predicted based on the summary of the kurtosis values for the 27 experiments shown in Fig. 3.3. For example, the tool in Experiment 9 should have the worst tool condition because it suffered the highest impact during machining. Meanwhile, the tool in Experiment 4 can be classified as in good condition because it encountered by low impact owing to the support of its optimum tool geometries during drilling activities. As mentioned in the previous section, the criteria of tool conditions are set based on the level of kurtosis values, which refer to the impact encountered by the twist drill tools. Therefore, several tool data from different experiments with different levels of kurtosis values were selected in this analysis to study their response to FFT and STFT. Tool data from Experiments 4, 9, 12, and 21 were selected for further analysis.

Figure 3.4 shows the FFT and STFT analyses based on the time response of the tool used in Experiment 4. The FFT figure shows no dominant frequency in the x and y axes. However, in the z-axis, two dominant frequencies at 700 and 200 Hz and a low FFT level are observed. In this analysis, the x and y axes, which are in balance position tool position, followed the basic rule of the drilling process.

Meanwhile, based on the spectrogram shown in the STFT analysis, the high impact of vibrations only happens during the first contact between the tool drill and the surface of the drilling material. The geometry parameters applied for the tool used in Experiment 4 managed to sustain the impact during the deep drilling activity. This tool presented the lowest kurtosis values overall for all regions, as shown in Fig. 3.3.

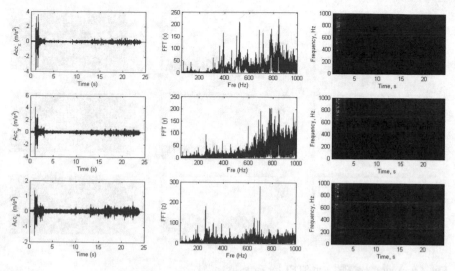

Fig. 3.4 Analysis of FFT and STFT for Experiment 4

Meanwhile, when the tool geometries of the point and helix angles increased and decreased by 5, as in Experiment 12, the capability of the tool to absorb impact from the drilling activity also decreased (Fig. 3.5). During the first 10 s, the in-depth drilling process runs as usual. At 14 s, the values of vibrations in the three axes begin to increase, but the process continues. The increase of vibration levels at 15 s due to the temporary chip clogging condition because the tool is still balanced position between the x and y axes, as shown by the vibration time response levels at 14–15 s.

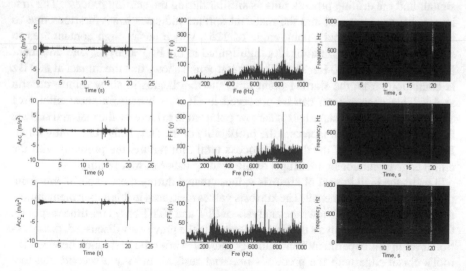

Fig. 3.5 Analysis of FFT and STFT for Experiment 12

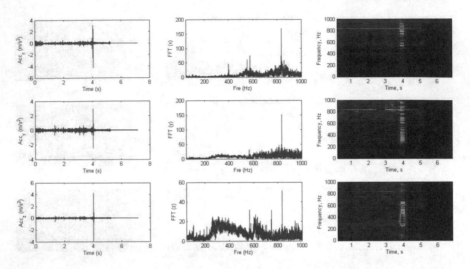

Fig. 3.6 Analysis of FFT and STFT for Experiment 21

At 20 s after a higher cut depth, the tool cannot sustain the drilling process effect. The chisel edge of the tool begins to blunt, thereby it's affecting the straightness of the tool drills. Meanwhile, all the dominant frequencies highlighted in the FFT analysis are referred to as the phenomena that happen at the 20 s. The tool loses its balances, as shown in the different vibrations in the x and y axes in the time domain.

Figure 3.3 shows that the tool drill in Experiment 21 has the highest kurtosis value in Region 3. It is also higher than the kurtosis values of the tools in Experiments 4 and 12. Figure 3.6 shows the response of the tool used in Experiment 21 on vibration signals and the drilling process runs as similar during the starting process. The first impact between the tool and the material surface indicates low vibrations due to the wide size of the tool's point angle of 125°, which has enough contact area to support the tool from vibration. As highlighted in the FFT analysis, one dominant frequency occurs at 800 Hz. The STFT spectrogram shows that the impact at 800 Hz is continuous from the starting point until the tool begins to show failure criteria at 4 s. This phenomenon can be observed in all axes because a substantial area contact between the chisel edges, a narrow point angle to increase the material surface friction. These situation increases the probability of the chisel edge becoming blunt. From the beginning of the drilling process until tool failure, the problem with the chisel edge continuously interrupts as the leading cause of tool failure.

Lastly, the high impact of kurtosis values created during these experiments from the tool used in Experiment 9. The kurtosis values measured in this experiment almost reach 1400. Figure 3.7 shows the analysis of FFT and STFT based on-time response. Compared with the analysis of other tools used in the previously discussed, the vibration level in this experiment continuously increased after the first touch between the tool's chisel edge with the product's material surface. In only a second, the tool

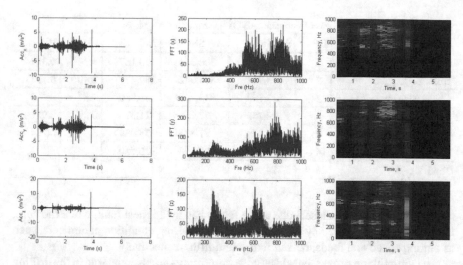

Fig. 3.7 Analysis of FFT and STFT for Experiment 9

began to stabilize. However, the vibration level during the drilling process continuously increased. The STFT spectrogram shows that the tool drills continuously encountered the impact. At 3 s, the tool already lost its balance, which refers to the blunt or broken chisel edge. At this time, the tools in the x and y axes are no longer in a balanced position. With the sharp point angle and the higher feeds and cutting speed, the cutting edge already broken upon its first contact with the material surface.

3.4.2 Analysis of Main Effects and ANOVA

The main effect analysis was obtained from the average kurtosis values of factors for each level. Table 3.2 shows the results of the responses of kurtosis values for each level of factors, namely, point angle, helix angle, clearance angle, feeds, and cutting speed in an integrated manner. The ranking was evaluated based on the differences of the highest and lowest averages of the kurtosis value for each factor. The generated

Table 3.2 Results for main effects from kurtosis value

Level	Point angle (°)	Helix angle (°)	Chisel edge angle (°)	Feeds (mm/rev)	Cutting speed (m/min)
1	534.8	479.7	459.2	351.6	441.3
2	315.4	179.0	146.2	383.2	313.1
3	307.1	498.6	552.0	422.5	402.9
Delta	227.7	319.6	405.8	71.0	128.2
Rank	3	2	1	5	4

Table 3.3 Results for analysis of variance (ANOVA) from kurtosis value

Source	DF	Seq SS	Adj SS	Adj MS	F	P
Point angle (°)	2	300,334	300,334	150,167	1.24	0.314
Helix angle (°)	2	578,942	578,942	289,471	2.40	0.123
Chisel edge angle (°)	2	813,760	813,760	406,880	3.37	0.060
Feeds (mm/rev)	2	22,755	22,755	11,378	0.09	0.911
Cutting speed (m/min)	2	77,878	77,878	38,939	0.32	0.729
Error	16	1,930,393	1,930,393	120,650		
Total	26	3,724,061				

ranking shows that the chisel edge angle factor has the highest rank, while the helix angle is second, followed by the point angle. The cutting condition parameters, such as cutting speed and feeds, ranked fourth and fifth, respectively.

The top-ranking proves that chisel edge angle plays an essential role in the drilling process because it controls the lip position, though the point angle does not have a sharp shape. As long as the clearance angle provides support to the lips and the heat generated by the process can be carried away, then this case is not a problem. This case happened in Experiments 22–24. In the second-ranking, the helix angle was only left behind with 21% of the delta percentage with clearance angle, showing that the helix angle can also contribute to the success of the drilling mechanism. Again, Experiments 22–24 proved that a wide helix angle and a low chisel edge angle could reduce the impact encountered during drilling, though the angles are shaped with a flared point angle. ANOVA will be discussed to examine the parameter again and validate the results shown in the analysis of mean effect.

Table 3.3 shows the ANOVA values generated by Minitab 16 statistical software. The influence of a factor on the quality characteristics can be examined through the generated value number. The result confirmed that the chisel edge angle, which has an F-ratio value of 3.37, as the most significant parameter. The helix angle is second followed by point angle with 2.40 and 1.24 F-ratio values, respectively. Meanwhile, the same as the mean effect analysis, cutting condition parameters only have a small effect on the performance of tool drills with cutting speed and feeds of 0.32 and 0.09, respectively. The effect of feeds is less than that of cutting speed because the value of the feed rate directly depends on the cutting speed.

3.5 Optimum Tool Selection

Based on the detailed discussion, the optimum tool should first require a balance chisel edge angle to reduce pressure and provide enough support to the drill's cutting lip. Second, the optimum tool should be formed by widening the helix angle to carry away chipping and create heat easily. Next, it should have a sharp shape to reduce the contact point between the chisel edge and the material surface. In terms of cutting

Table 3.4 Optimum twist tool drill geometries

Tool geometries	Optimum parameter
Point angle	115°
Helix angle	35°
Chisel angle	120°

conditions, as long as the optimum geometry parameters are applied, the tools can survive with 30–50 m/min of cutting speed and 0.1–0.3 mm/rev of feeds, as shown in Experiments 4 to 6. Even with a high feed rate value of more than 350 mm/min, the tool used in Experiment 6 maintained the impact and reduced the kurtosis value to below 200.

Meanwhile, the worst tool geometries are those used in Experiments 7 to 9, because even though Experiment 7 began with a low cutting speed of 30 m/min, the summation values of kurtosis are more than 500. Cutting speed increments of 40 and 50 m/min for Experiments 8 and 9 increased the summation of kurtosis by 65%. A simple comparison between the tool used in Experiment 5, which is represented as the excellent tool, and the tool used in Experiment 8, which is represented as the worst tool, can be seen in Fig. 3.4. The tools used in Experiments 5 and 8 were selected because they share the same cutting speed and the gaps in their kurtosis values were the lowest. The comparison of kurtosis values was too far for the z and y axes, where the kurtosis values in the z-axis for the tool used in Experiment 8 was only 22 compared to 645 for Tool 8. Next, the kurtosis value of Tool 5 for the vibration signal in the y-axis is 20, while that for Tool 8 was more than 280. Meanwhile, not much change happened in the x-axis. Thus, the same results obtained by Harun et al. [18], where the x-axis not much contribute to the deformation of the tool during the drilling process.

Therefore, the suggested optimal tool that can sustain the performance of tool drills is the tool used in Experiments 4 to 6 that has a sharp 115° point angle, a wide 35° helix angle, and an equal 120° clearance angle, as shown in Table 3.4. By using the optimised tool, the impact during the drilling process can be sustained, as shown in Fig. 3.8 for the main effect. It is shown that the lowest optimum parameters in four out of five of the proposed tools validated by the data shown in Fig. 3.3, which recorded low kurtosis values.

3.6 Discussions

Table 3.1 shows Experiments 1–9 in region 1 supplied the point angle's parameter at 115°. Based on the theory, a sharp point angle's tool drill lower cause lower tool wear compared to the wide point angle [4]. A lower contact between the material surface and the chisel edge can reduce the potential tool chisel edge from being blunt

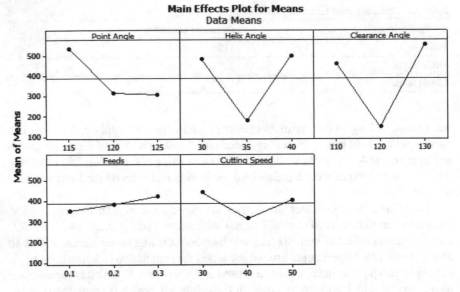

Fig. 3.8 Plot of the main effect of kurtosis value

or broken. Experiments 1 to 3 used similar parameters, except for cutting speed as tabulated in Table 3.1.

The tool drills applied in Experiments 1 to 3 have sharp point angles (115°), which accelerated cutting during the drilling process. However, with a narrow helix angle (30°), flowing out from the operation area was trying for a chip because the hole clogged the chips. The clogging problem worsened when the cutting speed increased from 30 to 50 m/min. In Experiment 3, the cutting process accelerated and the clogging impact of the chips encountered by the tool drills increased. Furthermore, pressure increased with a small clearance angle of 110°. This phenomenon can be seen in Fig. 3.3, where the kurtosis value increased from Experiments 1–3 by 91% when the cutting speed increased from 30 m/min to 50 m/min.

Meanwhile, different scenarios happened to the tool drills for Experiments 4–6. When the tools used the same point angle of 115° in Experiments 1–3, the 35° helix angle was more extensive than the previous tools with feeds increased by 0.2 mm/rev. Although the feed was higher than the previous experiment, the number of recorded kurtosis was lower than the previous experiment. The chips moved out more full with sufficient support from the lip thickness to carry away the heat generated with adequate clearance angle [10].

However, high clearance angle also causes drills to break down because of insufficient support of the lip, and lip thickness is insufficient to carry away the generated heat. In Experiments 7–9, the parameters for helix angle, clearance angle, and feeds were 40°, 130°, and 0.3 mm/rev, respectively. Figure 3.8 shows that the kurtosis numbers for Experiments 7–9 were higher than those of the previous experiment (4–6). However, the tool drills used were shaped by widening the helix and clearance

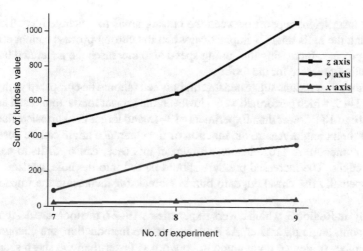

Fig. 3.9 Kurtosis values in each axis for Experiments 7–9

angles. Clearance angle plays a vital role in shaping a good tool drill. With a small drill point created by the sharp shape of the point angle, the high degree for clearance angle, and the high feed rate more than 450 mm/min, the chisel edge might break at the first touch between the tool's chisel edge and the material surface. As we have known, the chisel edge works as a guide that controls the straightening of the drill [14]. When the chisel edge breaks, the tool is no longer stable and loses its principle. Figure 3.8 shows the values of kurtosis for the feed axis. The kurtosis value in the z-axis is higher than the other axes owing to the high impact between the chisel edge and the material surface (Fig. 3.9).

Meanwhile, for Experiments 10–18, which is known as Region 2, the tools are designed with a point angle of 120°. Tools used in Experiments 10 to 12 are 30° helix angle, 120° clearance angle, and 0.3 mm/rev feeds. With an increased area contact of point angle and a narrow helix angle, this condition should be more complicated than the condition in Experiments 1–3. However, with an equal clearance angle and high numbers of feed rate, the cutting lips are supported, and the chips are pushed to move out from the drilling area. When the cutting speed increases, the amount of chips produced also increases, thereby hindering the tool drills from absorbing more impact. As a result, the chips are clogged, which is the same problem faced by Experiments 1–3. In summary, a high feed helps to push the chip flow out and balances the clearance angle with enough support to the cutting lips as created by Experiments 10–12 and Experiments 1–3.

Experiments 13–15 used tools with a wide helix angle (35°), long clearance angle (135°), and low feed rate. However, not much change can be observed the kurtosis values for Experiments 13–15 and almost similar to Experiment 11, except for Experiment 15. Although the increase of the helix angle allowed the chip to move out easily, slow feeds make it challenging to push chips out continuously [1]. Additionally, as explained in the previous paragraph, high clearance causes cutting lips to provide insufficient support between the cutting areas to generate heat. Therefore, expanding the helix angle is impractical when the cutting process cannot afford the cutting process. Increasing the cutting speed also increases the generated heat, thus contributing pressure for the tool.

Previous discussions suggested that higher feeds should be compared with Experiments 13–15, which performed well. However, the impact for the tools used in Experiments 16 to 18 is lower than Experiments 13–15 and larger the kurtosis values, even with 40° helix angle. Again, the function of the clearance angle can be observed in this phenomenon. A 110° clearance angle, in this case, causes drills to have little cutting edges. The increased pressure should be fed into the hole, causing the tool drill, especially the chisel edge, to blunt or break for increasing the impact of the drill [9].

Lastly, in Region 3, which covers Experiments 19–27, the tools are designed with a 125° point angle parameter. As explained in the Introduction, the change for the wide number of degree point angle to be worn is faster than the sharp shape point angle because of the high contact between the material surface and the tool point angle and chisel edge [8].

The scenario in Experiments 19–21 is almost the same occurred in Experiments 10–12. The large area of contact between the chisel edge and the material surface, and the high clearance angle and narrow helix angle seem to be contradicting. With flared shaped, the area of contact should be supported by low clearance, providing enough cutting edges to cut and carry away the generated heat. The wide helix angle should move the chips out rapidly to avoid the clogging process. Meanwhile, for Experiments 19–21, high clearance angle and narrow helix angle are not workable. The reason is that the cutting process disturbed by low support because of the cutting edge and a narrow helix angle causes chips to stick in the cutting area, which gives high pressure to the tool. The increasing cutting speed from 30 m/min to 50 m/min for Experiments 19 and 21 increased the friction between the chisel edge and the material surface. Consequently, incoming chips increased the impact on the tool causes the kurtosis value continuously increased with the increasing of cutting speed.

The problem encountered by Experiments 22–24 is similar to that in Experiments 21–23. The tools used by these experiments have 35° helix angle and 110° clearance angle. The feeds are 20 mm/rev lower than in the previous experiment with 0.3 mm/rev feeds. The drilling phenomena encountered by the tools in these experiments are similar to Experiments 21–23. However, the wide helix angle helps chips to flow out efficiently. A low clearance angle provides enough support to the cutting edges to cut even though the clearance is small and the feeds in feed axis (z) are high. The wide helix angle helps pull out chips from the cutting areas. Therefore, Fig. 3.7 shows that the kurtosis number is reduced when the cutting speed increases. Thus, the feed rate increase together with the cutting speed.

Lastly, for Experiments 25–27, not much difference was observed in terms of tool performance compared to previous experiments (22–24). With a balance clearance angle and a wide chip route, these experiments encountered lower impact than Experiments 22–24. However, low feeds (0.1 mm/rev) cause the chips to come out slowly from the cutting area, thus increasing the potential of chips to be stuck and place pressure on the tool drills. The impact encountered by the tool decreases with the kurtosis number when the feed rate increases.

Based on the detailed discussion in the previous section, each parameter function is recognized. However, each parameter function was evaluated in combination with other parameters and not individually. The author believes that evaluating the parameters individually contributes more impact during the drilling process. Moreover, combined evaluation helps for selecting the optimum parameter that contributes to reducing the impact of vibration during the drilling process. Therefore, by combining the analysis discussed earlier in several sections, such as the kurtosis level analysis of the main effect and ANOVA, the optimum parameters for the drilling process can be determined. Several studies implemented the Taguchi method as their parameter optimization method for different case studies and purposes. For example, Siddique et al. [19] implemented the Taguchi method to minimize the roughness of the product's surface during deep drilling activity. However, the results obtained a deficient feed; precisely, 0.04 mm/s and 25 mm of holes depth as the optimum cutting conditions for minimizing surface roughness. To the author's knowledge, it is not satisfied with the in-depth drilling process principle, which requires a high metal removal rate.

Furthermore, the depth of the hole of 25 mm was still in a shallow depth of cut. Meanwhile, Xavier and Elangovan [20] concluded that several factors should be considered to achieve the effectiveness of in-depth drilling process by optimising the tool geometry parameters. Similar to Sklenicka and Cesakava [7], they also studied the influence of geometry parameters on the magnitude of cutting forces. In this study, the suggested optimal tool can sustain the performance of tool drills with a sharp point angle of 115°, wide helix angle of 35°, and balanced chisel angle of 120°. This tool can carry out the drilling process with the lowest kurtosis values and the verified with frequency and time-frequency analyses.

3.7 Conclusion

In sum, kurtosis analysis indicates that the optimum tool geometry parameters with cutting parameters from tri-axial acceleration signal during the deep drilling process. ANOVA and main effect analysis showed that the most significant parameter affects the kurtosis value through the descending rank of clearance angle, helix angle, point angle, cutting speed, and feeds. The suggested optimal tool that can sustain the performance of tool drills is the tool used in Experiments 4 to 6, which has a sharp point angle of 115°, wide helix angle of 35°, and balanced chisel angle of 120°. This tool sustains impact during the drilling process with low kurtosis values and verified with signal processing techniques of frequency and time-frequency analyses. The performance of the tool drills in the drilling process can be improved through the application of kurtosis analysis, main effect, ANOVA and signal processing technique.

Acknowledgements The authors would like to thanks to Ministry of Science, Technology and Innovation (MOSTI) and Universiti Malaysia Pahang for providing laboratory facilities and financial assistance under e-science research fund project no. RDU140506.

References

1. H.M. Ertunc, K.A. Loparo, H. Ocak, Tool wear condition monitoring in drilling operations using hidden markov models. Int. J. Mach. Tools Manuf. **41**(9), 1363–1384 (2001)
2. US Industrial Outlook 1992: Business Forecasts for 350 Industries, US Department of Commerce, International Trade Administration (1992)
3. Deep Hole Drilling Applications Article. Available online at: http://www.unisig.com/gundri lling-education/deep-hole-drilling-applications.php
4. A. Uysal, M. Altan, E. Altan, Effects of cutting parameters on tool wear in drilling of polymer composite by Taguchi method. Int. J. Adv. Manuf. Technol. **58**, 915–921 (2012)
5. J. Mazoff, Drill pint geometry article, Malaysia retrieved December, 21 2016 Available online at: www.newmantools.com/machines/drillpoint.html
6. X. Wang, C. Huang, B. Zou, H. Liu, J. Wang, Effects of geometric structure of twist drill bits and cutting condition on tool life in drilling 42CrMo ultrahigh-strength steel. Int. J. Adv. Manuf. Technol. **64**, 41–47 (2013)
7. J. Sklenicka, I. Cesakava, Influence of the twist drill geometrical parameters on magnitude of cutting forces. MM Sci. J., 230–231 (2011)
8. G. Tosun, Statistical analysis of process parameters in drilling of AL/SICp metal matrix composite. Int. J. Adv. Manuf. Technol. **55**, 477–485 (2011)
9. A. Paul, S.G. Kapor, R.E. Devor, Chisel Edge and cutting lip shape optimization for improved twist drill point design. Int. J. Mach. Tools Manuf. **45**, 421–431 (2005)
10. M.H.S. Harun, M.A.N. Kamarizan, M.F. Ghazali, A.R. Yusoff, Statistical analysis of deep drilling process conditions using vibrations and force signals. MATEC Web Conf. **74**, 00002 (2015)
11. I. Abu-Mahfouz, Drilling wear detection and classification using vibration signals and artificial neural network. Int. J. Mach. Tools Manuf. **43**(7), 707–720 (2003)
12. C.R. Jochem, Mechanics and dynamics of drilling. Ph.D. Thesis, Faculty of Applied Science, University British Columbia (2006)
13. C. Malave, Deep hole drilling: Cutting forces and balance of tools. Master Thesis, Hogskolan I Gavle Academy (2015)
14. E.J.A. Armarego, R.H. Brown, *The Machining of Metals*, book edn. (N.J Prentice Hall, Englewood cliffs, 1969)
15. M.H.S. Harun, M.F. Ghazali, A.R. Yusoff, Tri-axial time-frequency analysis for tool failures detection in deep twist drilling process. Procedia CIRP **46**, 508–511 (2016)
16. X. Jie, Drill wear prediction and drilling conditions recognition with newly generated features. Ph.D. Theses, Hiroshima University (2014)
17. X. Li, S.K. Tso, S. Member, J. Wang, Real-time tool condition monitoring using wavelet transforms and fuzzy techniques. IEEE Transit. Sys. Man Cybern. Part C, Appl. Revision **30**(3), 352–357 (2000)
18. M.H.S. Harun, M.A.N. Kamarizan, M.F. Ghazali, A.R. Yusoff, Analysis of tri-axial force and vibration sensors for detection of failure criterion in deep twist drilling process. Int. J. Adv. Manuf. Technol. (2016)

19. A.N. Siddiquee, Z.A. Khan, P. Goel, M. Kumar, G. Agarwal, N.Z. Khan, Optimization of deep drilling process parameters of AISI 321 steel using Taguchi method. Procedia Mater. Sci. **6**, 1217–1225 (2014)
20. L.F. Xavier, D. Elangovan, Effective parameters for improving deep hole drilling process by conventional method—a review. Int. J. Eng. Res. Technol. (IJERT) **2**(3) (2013)

Chapter 4
Development of $Ti_{50}Ni_{50-X}Co_X$ (X = 1 and 5 at. %) Shape Memory Alloy and Investigation of Input Process Parameters of Wire Spark Discharge Machining

Hargovind Soni, S. Narendranath, M. R. Ramesh, Dumitru Nedelcu, P. Madindwa Mashinini, and Anil Kumar

Abstract The present study discusses, wire spark discharge machining and optimizations of its process parameters (Voltage (V), pulse off duration (T_{off}), pulse on duration (T_{on}), wire-speed (WS) and servo feed (SF)) for machining of $Ti_{50}Ni_{49}Co_1$ and $Ti_{50}Ni_{45}Co_5$ alloys conventionally used bone staple material. These alloys are developed through Vacuum arc melting (VAM) for the current investigation. For the elemental analysis, Energy Dispersive X-Ray Spectroscopy (EDX) investigations have been performed for the developed alloys. A hybrid combination of optimization techniques is used for parametric optimization in WSDM. The Grey relational analysis (GRA) Entropy measurement methods and response surface methodology (RSM) are used for formulating a hybrid combination of optimization techniques. Such type of technique is rarely implemented to get optimum machining conditions for wire spark discharge machining of TiNiCo alloy to achieve a higher metal removal rate with a better surface finish. Response surface design (DOE) is applied for an experimental plan and created L-33 orthogonal array. Machining has been carried out

The original version of this chapter was revised: The author's name was updated from "Madindwa Mashinin" to "P. Madindwa Mashinini". The correction to this chapter is available at https://doi.org/10.1007/978-3-030-50312-3_9

H. Soni (✉) · P. M. Mashinini
Mechanical and Industrial Engineering Technology, University of Johannesburg (Doornfontein Campus), Johannesburg, South Africa
e-mail: hargovindsoni2002@gmail.com

S. Narendranath · M. R. Ramesh
Department of Mechanical Engineering, National Institute of Technology Karnataka, Surathkal, India

D. Nedelcu
Department of Machine Manufacturing Technology, "GH ASACHI", Technical University of Iasi, Iasi, Romania

A. Kumar (✉)
Department of Mechanical Engineering, Delhi Technological University, New Delhi, India
e-mail: anilkumar76@dtu.ac.in

© Springer Nature Switzerland AG 2021, corrected publication 2021
S. Pathak (ed.), *Intelligent Manufacturing*, Materials Forming, Machining and Tribology, https://doi.org/10.1007/978-3-030-50312-3_4

of TiNiCo alloy as per the experimental plan and measured the machining responses (metal removal rate and average roughness). Experiment numbers 26 and 18 are found optimum machining conditions for $Ti_{50}Ni_{49}Co_1$ and $Ti_{50}Ni_{45}Co_5$ alloy, respectively, which is also confirmed by confirmation test results. Moreover, 6.52 mm^3/min higher metal removal rate has been noticed for the $Ti_{50}Ni_{49}Co_1$ and 8.538 mm^3/min metal removal rate for $Ti_{50}Ni_{45}Co_5$ alloy which means metal removal rate is increasing with increased the percentage of Co in TiNi which is also confirmed by scanning electron microscopy images. A similar trend has been observed for the average roughness.

Keywords Energy dispersive X-ray spectroscopy · Optimization techniques · TiNiCo alloy · Wire spark discharge machining

4.1 Introduction

NiTi-based smart materials are the unique class of shape memory material that regains the parent shape under proper thermal exposure. These alloys are most famous for the developments of biomedical and engineering instruments such as orthodontic arch-wires, actuators, stents, etc. because of their excellent properties [1, 2]. These alloys show outstanding properties such as shape memory effect, superelasticity, corrosion resistance, high specific strength, biocompatibility [2, 3]. The properties of these alloys are augmented by adding other elements (cobalt, silver, and copper, etc.) [4, 5]. The addition of cobalt in TiNi alloy improves the stability in properties for bio-implants applications [6, 7]. Many researchers have been studied on the influence of third alloying elements in TiNi based smart materials, but machining of ternary alloys and machining characteristics are very less. Conventional machining of smart materials is challenging since these materials are responsive to loading and operating conditions. Conventional machining techniques impose challenges while machining such smart materials. Therefore, unconventional machining techniques (WSDM, Abrasive water jet cutting, and Electrochemical cutting process) are suggested while machining these materials [7–11]. TiNiCo smart material was machined by wire spark discharge machining process which has varying elemental composition [12]. In the WSDM process machining, responses are influenced by cutting process parameters such as voltage, pulse on/off duration, the diameter of wire and wire speed. Spark erosion and vaporization cause material removal, which is influenced by voltage and pulse on/off durations [13]. Thermal stress, thick recast layer and irregular material removal have been noticed on the machined component. Machined surfaces through this technique have [14, 15]. These defects reduce the service life of the machine component due to residual stress [16–19]. In order to understand the effect of these process parameters, an experimental study was carried out at different combinations of input process parameters, and respective responses were studied to figure out the most suitable and optimized conditions [13, 18, 19]. The literature on machining of biomedical implant material using WEDM process is limited, which needs more studies on the analysis of the surface [20, 21]. Researchers have made many attempts using several optimization techniques. Response surface technique (RST) Grey and relational analysis (GRA) [22–25], and Optimum cutting parameters conditions for machining on D2 tool steel through spark discharge machining [26] have been used to optimize the cutting process parameters same as used for different machining. Process parameters have been optimized adopting hybrid grey relation analysis and entropy measurement technique during ultrasonic machining [27], machining of D2 tool steel through spark-discharge machining [26], wire spark discharge machining of EN 353 material [28], and wire spark discharge machining of WC-5.3%Co composite [29]. Techniques for Order Preferences by Similarity to Ideal Solution (TOPSIS) and Grey relational approach and have been used to optimize the milling process parameters [30]. Similarly, Taguchi based GRA was adopted for the optimization of turning process during magnesium alloy machining [31]. Even though several researchers have found optimum machining conditions WSDM using hybrid combination of optimization methods.

The literature review reveals that the Entropy-based GRA method with RSM is the most appropriate technique of optimization techniques to getting optimum machining conditions Wire spark discharge machining. Machining has been carried out of TiNiCo alloy as per the experimental plan (L-33 orthogonal array) and measured the machining responses (metal removal rate and average roughness). Machining process parameters will be optimized using hybrid combinations of optimization methods. Further study will be carried out with respect to confirmations test results.

4.2 Materials and Methods

4.2.1 Experimental Work

The detailed steps involved in the present work are shown in the flow chart represented in Fig. 4.1.

4.2.2 Melting of TiNiCo Shape Memory Alloys

TiNiCo alloys were developed using vacuum arc melting furnace have a purity of 99.4% Titanium, 99.89% Nickel and 99.49% Cobalt in compositions of $Ti_{50}Ni_{50-x}Co_x$ (x = 1 and 5 at.%). The vacuum was created 10^{-5} m bar inside the melting furnace after that argon gas into the chamber before the process of melting. The same process was carried out more than six times to confirm the homogeneity of advanced alloys. The shape of the developed material was in button form, as shown in Fig. 4.2. These buttons are afterwards sucked in the form 4 mm diameter and 80 mm long cylindrical rod. Details about the development of TiNiCo alloy have been presented by Soni et al. [32]. The vacuum arc melting setup used for the study (Fig. 4.2).

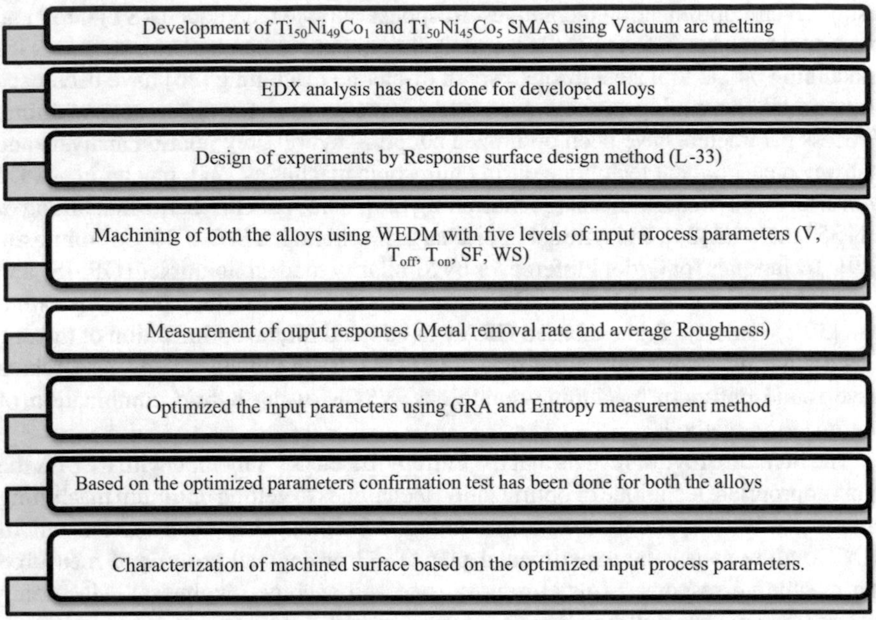

Development of $Ti_{50}Ni_{49}Co_1$ and $Ti_{50}Ni_{45}Co_5$ SMAs using Vacuum arc melting
EDX analysis has been done for developed alloys
Design of experiments by Response surface design method (L-33)
Machining of both the alloys using WEDM with five levels of input process parameters (V, T_{off}, T_{on}, SF, WS)
Measurement of ouput responses (Metal removal rate and average Roughness)
Optimized the input parameters using GRA and Entropy measurement method
Based on the optimized parameters confirmation test has been done for both the alloys
Characterization of machined surface based on the optimized input process parameters.

Fig. 4.1 Workflow chart adopted in the present work

Alloying elements
Buttons
Ammeter
Vacuum Arc Melting
Circular block
Copper crucible
Argon gas cylinder
Water Tank
TIG welding setup
Switch board
Cooling water supply
Rotary vacuum pump

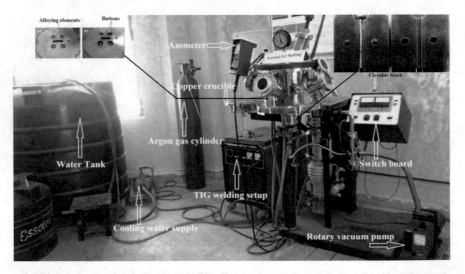

Fig. 4.2 Experimental setup of vacuum arc melting

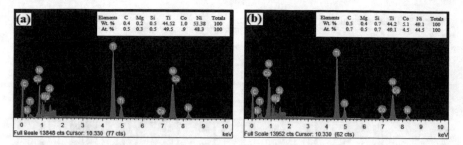

Fig. 4.3 EDX analysis of as-cast materials **a** $Ti_{50}Ni_{49}Co_1$ and **b** $Ti_{50}Ni_{45}Co_5$

4.2.3 EDX Analysis of As-Cast TiNiCo Alloys

To study the present elements of developed alloys, EDX analysis has been carried out through ZEISS, SEMV018 EDX spectrometer for both $Ti_{50}Ni_{49}Co_1$ and $Ti_{50}Ni_{45}Co_5$ alloys and found the satisfactory results as shown in Fig. 4.3a, b, respectively. Other elements such as Carbon, Silicon and Magnesium were also noticed during EDX analysis with the very less percentage in the developed alloys. These impurities were considered negligible due to a very less percentage.

4.2.4 WSDM of TiNiCo Alloys

WSDM (Electronica ELPULS15) used for $Ti_{50}Ni_{49}Co_1$ and $Ti_{50}Ni_{45}Co_5$ shape memory alloys machining. Voltage, T_{off}, T_{on}, WS, and SF are considered as input process parameters. Their effect on metal removal and average roughness was observed and recorded for further studies. Table 4.1 exhibits the considered machining process. Distilled water as the dielectric fluid and brass wire of diameter 0.25 mm used as a tool electrode for the machining operation. The machining process has been carried out by adopting the L-33 orthogonal array as a design of experiments that is generated by the response surface design approach (Table 4.2). Figure 4.4 shows the WSDM machining setup. The configuration of parts is a circular rod of 4 mm diameters and 80 mm length cross-sectioned using the WSDM process.

Table 4.1 Machining process parameters

S. No.	Input parameters	Units of selected parameters	Levels				
A	Voltage	(V)	20	30	40	50	60
B	Pulse off duration	(μs)	28	35	42	49	56
C	Pulse on duration	(μs)	105	110	115	120	125
D	Servo Feed	(mu)	2160	2170	2180	2190	2200
E	Wire Speed	(m/min)	2	3	4	5	6

Table 4.2 Experimental plan

S. No.	Coded					Uncoded				
	A	B	C	D	E	V	T_{off}	T_{on}	SF	WS
1	4	2	4	2	4	50	35	120	2170	5
2	4	4	2	4	2	50	49	110	2190	3
3	3	3	3	3	3	40	42	115	2180	4
4	3	3	3	3	1	40	42	115	2180	2
5	4	2	2	4	4	50	35	110	2190	5
6	2	4	2	2	2	30	49	110	2170	3
7	4	2	2	2	2	50	35	110	2170	3
8	4	4	4	2	2	50	49	120	2170	3
9	2	4	4	2	4	30	49	120	2170	5
10	4	4	4	4	4	50	49	120	2190	5
11	3	3	3	3	3	40	42	115	2180	4
12	4	4	2	2	4	50	49	110	2170	5
13	2	4	2	4	4	30	49	110	2190	5
14	2	2	4	4	4	30	35	120	2190	5
15	3	3	3	3	3	40	42	115	2180	4
16	2	2	4	2	2	30	35	120	2170	3
17	3	3	3	3	3	40	42	115	2180	4
18	2	3	3	3	3	30	42	115	2180	4
19	2	2	2	2	4	30	35	110	2170	5
20	2	4	4	4	2	30	49	120	2190	3
21	2	2	2	4	2	30	35	110	2190	3
22	4	2	4	4	2	50	35	120	2190	3
23	3	3	3	3	1	40	42	115	2180	2
24	5	3	3	3	3	60	42	115	2180	4
25	3	3	1	3	3	40	42	105	2180	4
26	3	3	3	1	3	40	42	115	2160	4
27	1	3	3	3	3	20	42	115	2180	4
28	3	3	3	3	5	40	42	115	2180	6
29	3	5	3	3	3	40	56	115	2180	4
30	3	3	3	3	2	40	42	115	2180	3
31	3	3	3	5	3	40	42	115	2200	4
32	3	1	3	3	3	40	28	115	2180	4
33	3	3	5	3	3	40	42	125	2180	4

Fig. 4.4   Machining setup of WSDM

Wire electrode

Clamps

Sparking Zone

Deionized water

SMA specimen

4.2.5   Measurements of Output Responses

4.2.5.1   Metal Removal Rate (MRR) and Average Roughness (AR)

The material removal rate is an important factor in the manufacturing industries because if the metal removal rate is high, then production is a large quantity of the product in lesser time will result in an efficient process. The metal removal rate can be calculated using Eq. (4.1) [33].

$$\text{Metal removal rate} = \text{Machining rate (mm/min)}$$
$$\times \text{ height of work material (mm)} \times \text{width of cut (mm)} \quad (4.1)$$

The cutting speed displayed in the machine is noted every definite time interval, and the average value of the cutting speed is used in the calculation of the metal removal rate. The width of the cut is measured in a scanning electron microscope image.

The average roughness (Ra) is another essential factor that influences product quality. Average roughness was investigated using "Tellysurf (Mitutoyo)" roughness tester with 3 mm evaluation length, 0.25 mm/s stylus rate and 0.8 mm cut off the length. The values of measured roughness are presented in Table 4.5.

4.2.6 Methodology

A hybrid combination of optimization techniques was used namely; grey relational technique, entropy measurement method and response surface design approach for both the alloys for obtaining the best combinations of process parameters. Figure 4.5 indicates the relation of all combination of optimization techniques.

4.2.6.1 Grey Relational Analysis (GRA)

GRA is one of the most practical analytical optimization methods. It provides suitable tools for examining the rank of an order of multiple objects with resemblance from an objective [34]. It also needs limited data to estimate the behaviour of an undefined system and discrete data problem. Factors are effaceable for broad sequence range. Therefore data pre-processing is a significant step to manage the factors of GRA [35]. The pre-processing data range varies from 0 to 1 [36]. Higher are lower are the better conditions and are considered for normalization in the present study for the

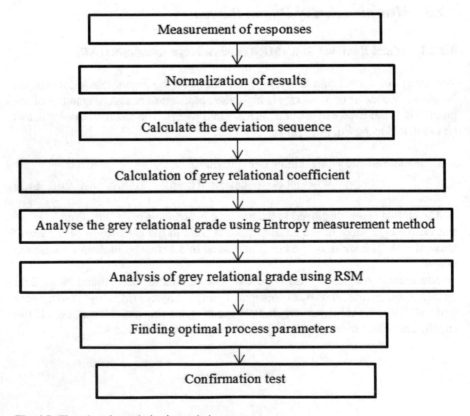

Fig. 4.5 Flowchart for optimization technique

Table 4.3 Conditions of normalization

S. No.	Conditions	Equations
1	Average roughness	$X_i^*(k) = \frac{X_i(k) - \min X_i(k)}{\max X_i(k) - \min X_i(k)}$ (4.2)
2	Metal removal rate	$X_i^*(k) = \frac{\max X_i(k) - X_i(k)}{\max X_i(k) - \min X_i(k)}$ (4.3)

Where, I = 1, 2n, k = 1, 2, y, p
$X_i^*(k)$ is the normalized value of the kth element in the ith sequence
Max $X_i(k)$ is the largest value of $X_i(k)$
Min $X_i(k)$ is the smallest value of $X_i(k)$
n is the number of experiments, and p is the number of quality characteristics

data pre-processing. The metal removal rate and roughness normalizations have been carried out with respect to higher and lower better, respectively. The same procedure will be considered for getting optimum results of both selected alloys. The detailed equations are presented in Table 4.3 [32].

After the normalization, the grey relational coefficient (γ) is calculated; this shows the interaction between optimal and actual normalized experimental results [29].

$$\gamma_i(k) = \gamma(x_0(k)) = \frac{\Delta \min + \zeta \Delta \max}{\Delta_{0,i}(k) + \zeta \Delta \max}.$$

$$i = 1; \dots; n; k = 1; \dots; p$$

(4.4)

where, $\Delta_{0,i}(k) = |x_0(k) - x_i(k)|$ is the difference of the absolute value called deviation sequence of the reference sequence $x_0(k)$ and comparability $x_i(k)$. The ξ is the different coefficient or identification coefficient. In general, it is set to 0.5. The GRG is a weighting-sum of the grey relational coefficients and defined as

$$\gamma(x_0, x_i) = \sum_n^{k=1} \beta_k(x_0, x_i)$$

(4.5)

where β_k represents the weighting value of the kth.

Performance characteristics, and $\sum_n^{k=1} \beta_k = 1$.

4.2.6.2   Entropy Measurement Technique

GRA determines the weights of each quality characteristics. The entropy technique is an objective weighting method. A discrete type of entropy is used in grey entropy measurement for properly conduct weighting analysis. Table 4.4 exhibits the necessary steps of the entropy measurement technique, which is used for calculating grey relational grade; there are six stages for calculating weights of each characteristic [37]. Grey relational grade (GRG) is estimated by multiplying the grey relational

Table 4.4 Basic steps for calculating grey relational grade

S. No.	Stages	Formulas
1	Summation of each attribute's value for all sequences Dk	$D_k = \sum\limits_{i=1}^{n} x_i(k)$ (4.6)
2	Calculate the normalization coefficient K where m is the number of attributes	$K = \frac{1}{(e^{0.5}-1)m}$ (4.7)
3	Specific attribute e_k to find the entropy	$e_k = \frac{1}{K} \sum\limits_{i=1}^{n} f\left(\frac{x_i(k)}{D_k}\right)$ (4.8)
4	Total entropy (E) calculation	$E = \sum\limits_{k=1}^{n} e_k$ (4.9)
5	weighting factor λ_k	$\lambda_k = \frac{(1-e_k)}{n-E}$ (4.10)
6	The normalized weight for each element can be calculated β_k	$\beta_k = \frac{\lambda_k}{\sum_{k=1}^{n} \lambda_i}$ (4.11)

coefficient with a respective weight of each characteristic, and the values for each alloy are given in Table 4.5.

4.2.6.3 Response Surface Technique

The response surface procedure examines the relation between selected parameters and one or more than one outputs [38]. Second-order polynomial equations used to in response surface approach. These equations are used to predict the value of experimental data as well as to find the optimum process parameters.

In order to study the effect of WSDM process parameters of $Ti_{50}Ni_{49}Co_1$, and $Ti_{50}Ni_{45}Co_5$ alloys on the average volumetric roughness and metal removal rate, a second-order polynomial response can be fitted into the Eq. (4.12) [39].

$$Y = \beta_0 + \sum_{i=1}^{k} \beta_i X_i + \sum_{i=1}^{k} \beta_{ii} X_i + \sum_{i,j=1,\ i\neq j}^{k} \beta_{ii} X_i X_j + \varepsilon \qquad (4.12)$$

where ε is the noise which is observed in the machining output Y. X_i is the linear input variables, X_i^2 and X_i, X_j is the squares and interaction terms, respectively, of these input machining parameters. The unknown second-order regression coefficients are β_o, β_i, β_{ij} and β_{ii}, which should be determined in the second-order model are obtained by the least square method.

The response surface approach was considered to design an experimental plan for appropriate and reliable measurement of the machining responses and creation of a mathematical model of second-order response surface with the best fittings. Thus produce a higher or lower value of response for optimum combinations of process parameters [40].

Table 4.5 Measured values of metal removal rate, average roughness and Grey relational grade (GRG)

S. No.	$Ti_{50}Ni_{49}Co_1$			$Ti_{50}Ni_{45}Co_5$		
	METAL removal rate (mm³/min)	SR (R_a)	GRG	Metal removal rate (mm³/min)	Average Surfaces (μm)	GRG
1	1.26	3.36	0.363	1.720	3.47	0.431
2	1.97	3.29	0.383	0.720	2.65	0.544
3	3.25	2.74	0.456	2.221	3.45	0.442
4	3.68	3.5	0.430	2.074	2.95	0.506
5	2.07	2.29	0.455	1.457	2.72	0.539
6	2.53	2.56	0.445	1.228	2.44	0.609
7	2.13	2.52	0.436	1.212	2.58	0.569
8	2.56	2.58	0.444	1.462	2.63	0.560
9	5.37	3.45	0.539	3.719	2.93	0.545
10	2.73	2.7	0.441	1.643	2.71	0.545
11	3.29	3.33	0.422	2.253	3.65	0.424
12	1.11	3.01	0.379	0.622	2.27	0.661
13	1.88	2.51	0.431	1.237	2.78	0.523
14	5.42	3.98	0.522	3.798	4.11	0.423
15	3.25	2.62	0.465	1.231	3.24	0.450
16	5.25	3.75	0.516	1.869	3.74	0.409
17	3.15	2.59	0.463	2.680	2.86	0.534
18	4.57	2.71	0.522	8.538	3.48	0.746
19	3.13	2.62	0.460	2.485	2.59	0.588
20	5.53	2.38	0.624	1.659	2.7	0.547
21	3.26	2.55	0.470	2.828	2.59	0.596
22	4.11	2.79	0.490	2.760	3.44	0.454
23	2.24	3.11	0.400	2.730	2.92	0.524
24	2.32	2.96	0.411	0.792	3.24	0.443
25	2.01	2.29	0.453	1.107	2.34	0.641
26	4.23	0.96	0.771	2.010	2.98	0.500
27	5.32	3.57	0.530	2.539	3.55	0.439
28	3.38	2.44	0.484	1.778	2.89	0.511
29	2.31	2.75	0.424	1.305	2.65	0.553
30	3.52	2.89	0.456	1.761	2.75	0.538
31	3.58	3.31	0.435	2.251	3.17	0.476
32	5.44	2.87	0.578	3.014	3.89	0.420
33	6.52	2.89	0.719	1.720	3.47	0.381

4.3   Results and Discussion

Most suitable input parameters are obtained through maximized values of grey relational grade (Table 4.5). The obtained GRG values were found different for both the alloys. Therefore the most suitable combination of process parameters obtained for these alloys is also different. According to GRG value, from Table 4.2, it was found that the inputs as mentioned in trial number 26 were most fitting for machining $Ti_{50}Ni_{49}Co_1$ alloy. Similarly, trial number 18 was optimized for $Ti_{50}Ni_{45}Co_5$ alloy.

4.3.1   Analysis of Metal Removal Rate

Figure 4.6a shows that the metal removal rate decreases with increased voltage (V) up to 50 V, and constant for $Ti_{50}Ni_{49}Co_1$ alloy. Initially, metal removal rate increases with voltage up to 30 V and further metal removal rate decreases for $Ti_{50}Ni_{45}Co_5$ alloy (Fig. 4.6b). The metal removal rate decreases at higher voltage as an increment in voltage are responsible for greater spark gap which decreases the strength of spark results in less amount of materials are removed from the surface of machined work material (Fig. 4.6a and 4.6b). Same as investigated for $Ti_{50}Ni_{40}Cu_{10}$ alloys during WSDM by others also [3]. Figure 4.6a, b also depict the metal removal rate reductions with increased T_{off} for both $Ti_{50}Ni_{49}Co_1$ and $Ti_{50}Ni_{45}Co_5$ alloys. The reduction in metal removal at higher T_{off} is due to the lesser intensity of spark was noticed, less intensity of sparks is not capable to removed more amount of material from the surface of work material comparatively. Metal removal rate increases with an increase in T_{on}, as shown in Fig. 4.6a for $Ti_{50}Ni_{49}Co_1$ alloy. A similar trend was seen in Fig. 4.6b up to 115 µs of T_{on}. The further small reduction was noticed in metal removal rate with an increased T_{on}, for $Ti_{50}Ni_{45}Co_5$ which may have occurred due to

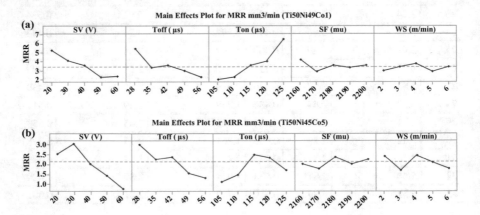

Fig. 4.6 Main effects plots for metal removal rate and process parameters for different alloys

poor flushing in the machining zone. Metal removal rate increases with increases in T_{on} because at high T_{on}, spark intensity is high. Because of the high intensity of spark, more thermal energy has been created in the machining zone, which can be removed more amount of material during the machining from the surface of work material. Hence, it is leading to a high metal removal rate. It is also noticed that the metal removal rate decreases up to 2170 (mu) servo feed (SF) and it increases fan from 2170 (mu) to 2180 (mu) further it was constant for $Ti_{50}Ni_{49}Co_1$ alloy. Similar trends noticed in Fig. 4.6b up to 2180 (mu) and again metal removal rate increases up to 2190 (mu) servo feed, further again decrease for $Ti_{50}Ni_{45}Co_5$ alloy. Hence SF has not much effect on metal removal rate since servo feed was not exclusive for these alloys. Other researchers report similar results during the machining of nanostructured hard facing materials by WSDM [41, 42]. Metal removal rate increases with an increase in WS up to 4 m/min. However, further, an increase in wire speed up to 5 m/min leads to lower the metal removal rate. A small improvement was noticed in metal removal rate with increased wire speed from 5 m/min to 6 m/min for $Ti_{50}Ni_{49}Co_1$ alloy, while in Fig. 4.6b initially metal removal rate decreases with wire speed up to 3 m/min and further metal removal rate increases with increased wire speed up to 4 m/min. Later metal removal rate decreases with increases in wire speed for $Ti_{50}Ni_{45}Co_5$ alloy. Till 4 m/min wire speed the value of metal removal was noticed high after that small reduction was noticed at 5 m/min wire speed then again increment has been noticed in the value of metal removal this due to vibration of wire during the machining zone which is also responsible for stability of the wire and results in not found sparking properly which leads to lower metal removal rate. Similar results have been noticed for Inconel 706 by using WSDM [43]. Effects plots for metal removal rate and SR are obtained by using MiniTab17 software.

4.3.2 Analysis of SR

From Fig. 4.7a, b, it was noticed that average roughness (SR) decreases with increases in voltage (V) up to 50 V. Further average roughness slightly increases up to 60 V for both $Ti_{50}Ni_{49}Co_1$ and $Ti_{50}Ni_{45}Co_5$ alloys. At high voltage, low average roughness is recorded due to less unwanted material is removed from the work surface; which can be cleaned easily from the machined surface through de-ionized water which is used as dielectric fluid. It is leading to low average roughness for both the alloys. It is clearly seen from Fig. 4.7a that average roughness slightly increased with increasing T_{off} up to 35 (μs) which is negligible and further average roughness decreases with all specific levels of voltage for $Ti_{50}Ni_{49}Co_1$ alloy. Similar trends have been noticed in Fig. 4.7b that average roughness decreases with increase T_{off} for $Ti_{50}Ni_{45}Co_5$ alloy. This is due to most of the amount has removed from the surface of work material at higher T_{off} through the flushing during the machining zone. Moreover, higher pulse off time also responsible for reducing the formation of micro globules and craters hence leading to low SR for both the alloys. Figure 4.7a indicates that average roughness increased with the improvement in T_{on} for $Ti_{50}Ni_{49}Co_1$ alloy but

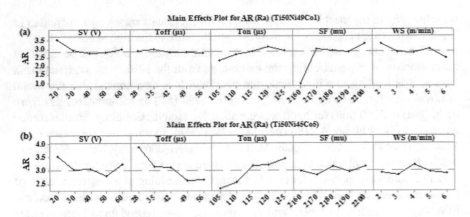

Fig. 4.7 Main effect plots for SR for different alloys

in Fig. 4.7b average roughness increases with an increased T_{on} up to 115 μs, and further, it slightly reduces for $Ti_{50}Ni_{45}Co_5$ alloy. This is due to the poor sparking during the machining process. Average roughness increases with an increased T_{on} because when it removes more quantity of material, some part of the melted material is cleaned by dielectric fluid and rest of the molten metal resolidify on the surface of the machined component. It forms craters and micro globules, leading to higher average roughness. It is observed that at low servo feed (SF) average roughness was low, but with the increase is servo feed up to 2170 mu, it was almost constant on other specific levels of servo feed because servo feed is related to the voltage. Usually, the spark rate will be 10–100 kHz which is very high, so servo feed control is limited for proper control of WSDM because it may result in loss of stability. This may cause server instabilities in the controlled feed rate leading to high average roughness for $Ti_{50}Ni_{49}CO_1$ alloy. On the other hand, it can be seen in Fig. 4.7b that there are not many effects of SR on SF for the machining of $Ti_{50}Ni_{45}Co_5$ alloy, which is the same as others noticed for $Ti_{50}Ni_{40}Cu_{10}$ alloys [44]. Figure 4.7a indicates that initially, average roughness was low at 2 m/min wire speed (WS), but later it decreases with the increased wire speed up to 6 m/min for the machining of $Ti_{50}Ni_{49}Co_1$ alloy. But in Fig. 4.7b it is seen that average roughness decreases with wire speed up to 3 m/min, and when wire-speed raised to 4 m/min, the reduction is noted in average roughness for the machining of the $Ti_{50}Ni_{45}Co_5$ alloy. Because at higher wire speed up to 6 m/min, average roughness reduces due to enhanced splashing of molten material. However, because of wire instability, irregular craters and also uneven spark generation occurs on the work surface [45].

4.3.3 Analysis of Microstructure

The microstructure of the machined surface of $Ti_{50}Ni_{49}Co_1$ and $Ti_{50}Ni_{45}Co_5$ alloys was studied from the image received through a scanning electron microscope (SEM) (ZEISS, product type- SEMV018 and product Serial number EVO-18-15-57) at different magnifications. Figure 4.8a, c represents the SEM images of the machined surface of the $Ti_{50}Ni_{49}Co_1$ alloy at the optimized process parameters (voltage of 40 V, T_{off} 42 μs, T_{on} 115 μs, SF 2160 mu and WS 4 m/min). Similarly Fig. 4.8b, c represents the SEM images for $Ti_{50}Ni_{45}Co_5$ alloy at the optimized process parameters (voltage of 30 V, T_{off} 42 μs, T_{on} 115 μs, SF 2180 mu and WS 4 m/min). These microstructures show cracks, craters, micro globules, microvoids and melted debris on the machined surface for both the alloys. Figure 4.8a, b are taken at low magnification; it is seen in Fig. 4.8a that the machining surface of $Ti_{50}Ni_{49}Co_1$ alloy exhibits low average roughness while Fig. 4.8b shows a high average roughness for $Ti_{50}Ni_{45}Co_5$ alloy at optimized process parameters. This is due to the effect of cobalt percentage in TiNi alloys and low voltage which is described in the analysis of average roughness that at low voltage average roughness will be high. It is evident from Fig. 4.8c, d that formations of micro globules are inevitable due to molten material resolidification due to dielectric quenching. Similar results were reported elsewhere on $Ti_{50}Ni_{35}Zr_{15}$ and

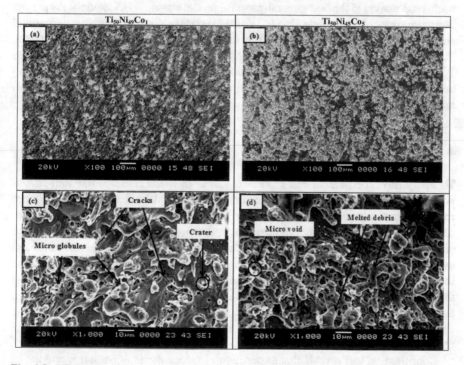

Fig. 4.8 SEM images of the machined surface for optimal process parameters. (i) $Ti_{50}Ni_{49}Co_1$ (**a** and **c**) (ii) $Ti_{50}Ni_{45}Co_5$ (**b** and **d**)

$Ti_{50}Ni_{49.5}Cr_{0.5}$ shape memory alloys using WSDM [5]. The formation of microvoids and micro globules at the given set of parameters are more prominent at a higher value of T_{on} as depicted in Fig. 4.8c, d. It is because thermal energy transferred to the material also increases, which results in higher material melting, leading to a rougher surface. It is evident from the crater size formed on the machined surface of the components. Similarly, at low voltage (30 V), micro globules and microvoids are more prominent, which tend to reduce at high voltage (40 V). The reason behind this is the increased gap voltage which increases the spark gap. Therefore facilitates better flushing and decreases the growth of microvoids and micro globules.

4.3.4   Prediction of WSDM Responses

ANN is a synthetic illustration of the human brain which simulates the learning approach using artificial neurons governed by mathematical models for processing of information. Neural networks are non-linear mapping systems. It consists of simple processors which are called neurons and linked by weighted relations. Where all neuron has inputs and generates an output. It can be seen as the reflection of local information that is stored in connections. The feed-forward neural network based on back-propagation is the best general-purpose model among ANN models. As per the literature survey for the perfection of WSDM responses through ANN, experimental results have been divided into two kinds of data, namely test data and training data. To get the best prediction values of responses, the training data should be considered 70% of experimental results and the rest of the data should be test data [46]. The details about the ANN prediction method author has been published in his previous article [47].

4.3.4.1   Confirmation Test Result

To verify the experimental and predicted values of metal removal rate and roughness confirmation test had been conducted for both the alloys based on the optimum machining cutting conditions are presented in Table 4.6. Experimental values of material removal rate and average roughness for both the alloys are given in Table 4.6. The small variation is observed between the predicted values of metal removal rate and average roughness at optimal process parameters during the confirmation test, and the difference is minimum. An artificial neural network (ANN) is used for predicting the responses. The author already published details of ANN in his previous research paper [47]. The reason for the minimum variation is; vibration of the wire during machining, poor sparking, less flushing and degradation of dielectric fluid in the machining area.

Confirmation test reveals that experimental values are in accordance and are in excellent agreement with the predicted values where the expected error is between 10 and 6% for the metal removal rate and 11–8% for average roughness for both

Table 4.6 Confirmation test result

Machined Alloys	Optimum machining conditions				Predicted responses		Experimental responses		Error (%)	
	V	T$_{off}$	T$_{on}$	SF	WS	Metal removal rate	Average roughness	Metal removal rate	Average roughness	Metal removal rate	Average roughness
	(V)	(μs)	(μs)	(mu)	(m/min)	(mm^3/min)	(μm)	(mm^3/min)	(μm)	(mm^3/min)	(μm)
Ti$_{50}$Ni$_{49}$Co$_1$	40	42	115	2160	4	4.23	0.96	4.65	1.06	10	11
Ti$_{50}$Ni$_{50}$Co$_5$	30	42	115	2180	4	9.01	3.48	8.76	3.75	6	8

the alloys respectively. Therefore, this method can be termed efficient in the case of machining of TiNiCo ternary shape memory alloys.

4.4 Conclusion

The metal removal rate and SR of $Ti_{50}Ni_{49}Co_1$ and $Ti_{50}Ni_{45}Co_5$ ternary SMAs in the WSDM process significantly relate to the inputs parameters of WSDM. The experiments were designed as per L-33 orthogonal array using a response surface design method to explore the effects of machining parameters on the output responses of the WSDM process. The principal conclusions are drawn as follows:

- Voltage of 40 V, T_{off} 42 μs, T_{on} 115 μs, SF 2160 mu and WS 4 m/min were found as the optimum combination of machining input process parameters for $Ti_{50}Ni_{49}Co_1$ alloy and machining has been carried out on these process parameters, average roughness 0.96 μm and 4.23 metal removal rate mm^3/min were achieved for $Ti_{50}Ni_{49}Co_1$ alloy. Similarly, for $Ti_{50}Ni_{45}Co_5$ alloy, average roughness 3.48 μm and metal removal rate, 8.53 mm^3/min were achieved at the best combination of input process parameters. The voltage of 30 V, T_{off} 42 μs, T_{on} 115 μs, servo feed 2180 mu and wire-speed 4 m/min were found as the optimum combination of machining input process parameters for $Ti_{50}Ni_{45}Co_5$ alloy.
- A hybrid combination of optimization techniques was effectively employed in the WSDM process to find the optimum combination of input process parameters. There is some variation in the predicted results at the same input process parameters for both the alloys. This may be due to a machining error during the confirmation test.
- Three process parameters, namely V, $T_{off,}$ and T_{on} are the most influencing machining parameters for metal removal rate and roughness for both the alloys. The rest of the parameters are not much influential metal removal rate and SR.
- SEM confirms the craters, micro globules, melted debris microvoid and cracks on the surface of work material which is machined at the optimum machining process parameters for both the alloys.

References

1. Z. Lekston, D. Stroz, M.J. Drusik-Pawlowska, Preparation and characterization of nitinol bone staples for cranio-maxillofacial surgery. J. Mater. Eng. Perform. **21**, 2650–2656 (2012). https://doi.org/10.1007/s11665-012-0372-3
2. T. Nam, T. Saburi, K. Shimizu, Cu-Content Dependence of Shape Memory Characteristics in Ti-Ni-Cu Alloys. Mater. Trans., JIM **31**, 959–967 (1990). https://doi.org/10.2320/matertrans1989.31.959

3. M. Manjaiah, S. Narendranath, S. Basavarajappa, V.N. Gaitonde, Effect of electrode material in wire electro discharge machining characteristics of $Ti_{50}Ni_{50-x}Cu_x$ shape memory alloy. Precis. Eng. **41**, 68–77 (2015). https://doi.org/10.1016/j.precisioneng.2015.01.008

4. L. Isola, P. La Roca, A. Roatta, P. Vermaut, L. Jordan, P. Ochin, J. Malarría, Load-biased martensitic transformation strain of $Ti_{50}–Ni_{47}–Co_3$ strip obtained by a twin-roll casting technique. Mater. Sci. Eng. A. **597**, 245–252 (2014). https://doi.org/10.1016/j.msea.2013.12.102

5. S.F. Hsieh, S.L. Chen, H.C. Lin, M.H. Lin, S.Y. Chiou, The machining characteristics and shape recovery ability of Ti-Ni-X (X=Zr, Cr) ternary shape memory alloys using the wire electro-discharge machining. Int. J. Mach. Tools Manuf. **49**, 509–514 (2009). https://doi.org/10.1016/j.ijmachtools.2008.12.013

6. R-r Jing, F-s Liu, The influence of Co addition on phase transformation behavior and mechanical properties of TiNi alloys. Chin. J. Aeronaut. **20**, 153–156 (2007)

7. A. Fasching, D.W. Norwich, T. Geiser, G.W. Paul, An evaluation of a NiTiCo alloy and its suitability for medical device applications. J. Mater. Eng. Perform. **20**, 641–645 (2011). https://doi.org/10.1007/s11665-011-9845-z

8. H.C. Lin, K.M. Lin, Y.S. Chen, C.L. Chu, The wire electro-discharge machining characteristics of Fe-30Mn-6Si and Fe-30Mn-6Si-5Cr shape memory alloys. J. Mater. Process. Technol. **161**, 435–439 (2005). https://doi.org/10.1016/j.jmatprotec.2004.07.079

9. S. Chun, J. Noh, J. Yeom, J. Kim, T. Nam, Intermetallics martensitic transformation behavior of Ti e Ni e Ag alloys. Intermetallics **46**, 91–96 (2014). https://doi.org/10.1016/j.intermet.2013.11.001

10. P.M. Mashinini, H. Soni, K. Gupta, Investigation on dry machining of stainless steel 316 using textured tungsten carbide tools. Mater. Res. Express **7** (2020). https://doi.org/10.1088/2053-1591/ab5630

11. H. Soni, P.M. Mashinini, Wire electro spark machining and characterization studies on $Ti_{50}Ni_{49}Co_1$, $Ti_{50}Ni_{45}Co_5$ and $Ti_{50}Ni_{40}Co_{10}$ alloys. Mater. Res. Express 7, 1–7 (2020). https://doi.org/10.1088/2053-1591/ab6196

12. S.F. Hsieh, A.W.J. Hsue, S.L. Chen, M.H. Lin, K.L. Ou, P.L. Mao, EDM surface characteristics and shape recovery ability of $Ti_{35.5}Ni_{48.5}Zr_{16}$ and $Ni_{60}Al_{24.5}Fe_{15.5}$ ternary shape memory alloys. J. Alloys Compd. **571**, 63–68 (2013). https://doi.org/10.1016/j.jallcom.2013.03.111

13. S. Kumar, M.A. Khan, B. Muralidharan, Processing of titanium-based human implant material using wire EDM. Mater. Manuf. Process. **34**, 695–700 (2019). https://doi.org/10.1080/10426914.2019.1566609

14. B. Puri, B. Bhattacharyya, Modeling and analysis of white layer depth in a wire-cut EDM process through response surface methodology. Int. J. Adv. Manuf. Technol. **25**, 301–307 (2005). https://doi.org/10.1007/s00170-003-2045-8

15. R. Rao, V. Yadava, Multi-objective optimization of Nd:YAG laser cutting of thin superalloy sheet using grey relational analysis with entropy measurement. Opt. Laser Technol. **41**, 922–930 (2009). https://doi.org/10.1016/j.optlastec.2009.03.008

16. M.M. Dhobe, I.K. Chopde, C.L. Gogte, Investigations on surface characteristics of heat treated tool steel after wire electro-discharge machining. Mater. Manuf. Process. **28**, 1143–1146 (2013). https://doi.org/10.1080/10426914.2013.822976

17. A. Kumar, V. Kumar, J. Kumar, Surface crack density and recast layer thickness analysis in WEDM process through response surface methodology. Mach. Sci. Technol. **20**, 201–230 (2016). https://doi.org/10.1080/10910344.2016.1165835

18. S. Daneshmand, V. Monfared, A.A. Lotfi Neyestanak, Effect of tool rotational and Al_2O_3 powder in electro discharge machining characteristics of NiTi-60 shape memory alloy. Silicon, 1–11 (2016). https://doi.org/10.1007/s12633-016-9412-1

19. P. Sharma, D. Chakradhar, S. Narendranath, Analysis and optimization of WEDM performance characteristics of Inconel 706 for aerospace application. Silicon, 1–10 (2017). https://doi.org/10.1007/s12633-017-9549-6

20. F. Kara, Taguchi optimization of surface roughness and flank wear during the turning of DIN 1.2344 tool steel. Mater. Test. **59**, 903–908 (2017). https://doi.org/10.3139/120.111085

21. F. Kara, B. Öztürk, Comparison and optimization of PVD and CVD method on surface rough-ness and flank wear in hard-machining of DIN 1.2738 mold steel. Sens. Rev. **39**, 24–33 (2019). https://doi.org/10.1108/sr-12-2017-0266

22. S.R. Elsen, T. Ramesh, Optimization to develop multiple response hardness and compressive strength of zirconia reinforced alumina by using RSM and GRA. Int. J. Refract. Met. Hard Mater. **52**, 159–164 (2015). https://doi.org/10.1016/j.ijrmhm.2015.06.007

23. Q. Yang, Y. Zhong, H. Zhong, X. Li, W. Du, X. Li, R. Chen, G. Zeng, A novel pretreatment process of mature landfill leachate with ultrasonic activated persulfate: optimization using integrated Taguchi method and response surface methodology. Process Saf. Environ. Prot. **98**, 268–275 (2015). https://doi.org/10.1016/j.psep.2015.08.009

24. S. Dewangan, S. Gangopadhyay, C.K. Biswas, Multi-response optimization of surface integrity characteristics of EDM process using grey-fuzzy logic-based hybrid approach. Eng. Sci. Technol. an Int. J. **18**, 361–368 (2015). https://doi.org/10.1016/j.jestch.2015.01.009

25. C. Raju, C. Sathiya Narayanan, Application of a hybrid optimization technique in a multiple sheet single point incremental forming process. Measurement. **78**, 296–308 (2016). https://doi.org/10.1016/j.measurement.2015.10.025

26. M.K. Pradhan, Optimization of MRR, TWR and surface roughness of EDMed D2 Steel using an integrated approach of RSM, GRA and entropy measurement method, in *2013 International Conference on Energy Efficient Technologies for Sustainability,* ICEETS 2013, pp. 865–869. (2013). https://doi.org/10.1109/ICEETS.2013.6533499

27. G.K. Dhuria, R. Singh, A. Batish, Application of a hybrid Taguchi-entropy weight-based GRA method to optimize and neural network approach to predict the machining responses in ultra-sonic machining of Ti–6Al–4V. J. Brazilian Soc. Mech. Sci. Eng. (2016). https://doi.org/10.1007/s40430-016-0627-2

28. A. Varun, N. Venkaiah, Simultaneous optimization of WEDM responses using grey relational analysis coupled with genetic algorithm while machining EN 353. Int. J. Adv. Manuf. Technol. **76**, 675–690 (2014). https://doi.org/10.1007/s00170-014-6198-4

29. K. Jangra, S. Grover, A. Aggarwal, Optimization of multi machining characteristics in WEDM of WC-5.3%Co composite using integrated approach of Taguchi, GRA and entropy method. Front. Mech. Eng. **7**, 288–299 (2012). https://doi.org/10.1007/s11465-012-0333-4

30. P.M. Gopal, K. Soorya Prakash, Minimization of cutting force, temperature and surface rough-ness through GRA, TOPSIS and Taguchi techniques in end milling of Mg hybrid MMC. Meas. J. Int. Meas. Confed. **116**, 178–192 (2018). https://doi.org/10.1016/j.measurement.2017.11.011

31. R. Viswanathan, S. Ramesh, V. Subburam, Measurement and optimization of performance characteristics in turning of Mg alloy under dry and MQL conditions. Meas. J. Int. Meas. Confed. **120**, 107–113 (2018). https://doi.org/10.1016/j.measurement.2018.02.018

32. H. Soni, S. Narendranath, M.R. Ramesh, Effect of machining parameters on wire electro discharge machining of shape memory alloys analyzed using grey entropy method. J. Mater. Sci. Mech. Eng. **2**, 50–54 (2015)

33. H. Soni, N. Sannayellappa, R. Motagondanahalli Rangarasaiah, An experimental study of influence of wire electro discharge machining parameters on surface integrity of TiNiCo shape memory alloy. J. Mater. Res., 1–9 (2017). https://doi.org/10.1557/jmr.2017.137

34. S.K. Majhi, M.K. Pradhan, H. Soni, Application of integrated RSM-grey-entropy analysis for optimization of EDM parameters, pp. 4–9 (2013)

35. R.K. Pandey, S.S. Panda, Optimization of bone drilling parameters using grey-based fuzzy algorithm. Measurement **47**, 386–392 (2014). https://doi.org/10.1016/j.measurement.2013.09.007

36. C. Raju, C.S. Narayanan, Application of a hybrid optimization technique in a multiple sheet single point incremental forming process. Measurement **78**, 296–308 (2016). https://doi.org/10.1016/j.measurement.2015.10.025

37. T. Mineta, T. Deguchi, E. Makino, T. Kawashima, T. Shibata, Fabrication of cylindrical micro actuator by etching of TiNiCu shape memory alloy tube. Sen. Actuat. A Phys. **165**, 392–398 (2011). https://doi.org/10.1016/j.sna.2010.12.002

38. M.S. Hewidy, T.A. El-Taweel, M.F. El-Safty, Modelling the machining parameters of wire electrical discharge machining of Inconel 601 using RSM. J. Mater. Process. Technol. **169**, 328–336 (2005). https://doi.org/10.1016/j.jmatprotec.2005.04.078

39. M.W.J. Layard, Institute of Mathematical Statistics is collaborating with JSTOR to digitize, preserve, and extend access to The Annals of Mathematical Statistics. ® www.jstor.org (n.d.)

40. A. Helth, U. Siegel, U. Kühn, T. Gemming, W. Gruner, S. Oswald, T. Marr, J. Freudenberger, J. Scharnweber, C.-G. Oertel, W. Skrotzki, L. Schultz, J. Eckert, Influence of boron and oxygen on the microstructure and mechanical properties of high-strength Ti$_{66}$Nb$_{13}$Cu$_8$Ni$_{6.8}$Al$_{6.2}$ alloys. Acta Mater. **61**, 3324–3334 (2013). https://doi.org/10.1016/j.actamat.2013.02.022

41. A. Saha, S.C. Mondal, Multi-objective optimization in WEDM process of nanostructured hardfacing materials through hybrid techniques. Measurement **94**, 46–59 (2016). https://doi.org/10.1016/j.measurement.2016.07.087

42. A. Saha, S.C. Mondal, Experimental investigation and modelling of WEDM process for machining nano-structured hardfacing material. J. Brazilian Soc. Mech. Sci. Eng. (2016). https://doi.org/10.1007/s40430-016-0608-5

43. P. Sharma, D. Chakradhar, S. Narendranath, Evaluation of WEDM performance characteristics of Inconel 706 for turbine disk application. Mater. Des. **88**, 558–566 (2015). https://doi.org/10.1016/j.matdes.2015.09.036

44. M. Manjaiah, S. Narendranath, S. Basavarajappa, V.N. Gaitonde, Wire electric discharge machining characteristics of titanium nickel shape memory alloy. Trans. Nonferrous Met. Soc. China (Engl. Ed.) **24**, 3201–3209 (2014). https://doi.org/10.1016/s1003-6326(14)63461-0

45. M. Manjaiah, S. Narendranath, S. Basavarajappa, Wire electro discharge Machining performance of TiNiCu shape Memory alloy. Silicon (2015). https://doi.org/10.1007/s12633-014-9273-4

46. G. Ugrasen, H.V. Ravindra, G.V.N. Prakash, R. Keshavamurthy, Estimation of machining performances Using MRA, GMDH and artificial neural network in wire EDM of EN-31. Procedia Mater. Sci. **6**, 1788–1797 (2014). https://doi.org/10.1016/j.mspro.2014.07.209

47. H. Soni, S. Narendranath, M.R. Ramesh, ANN and RSM modeling methods for predicting material removal rate and surface roughness during WEDM of Ti 50 Ni 40 Co 10 shape memory alloy. AMSE J. IIETA Adv. A. **54**, 435–443 (2018)

Chapter 5
Application Potential of Fuzzy Embedded TOPSIS Approach to Solve MCDM Based Problems

Akula Siva Bhaskar, Akhtar Khan, and Situ Rani Patre

Abstract The current chapter discusses the application potential of a multiple criteria decision making (MCDM) method in fuzzy environment. The hybridization of a MCMD-based method namely TOPSIS with Fuzzy set theory, has been proposed for solving a broad variety of problems associated with research and academic community. The concept of Multi-criteria decision making (MCDM) has been extensively adapted by several analysts in distinctive range of study. The amalgamation of MCDM with fuzzy interface has driven to an advanced decision theory called fuzzy multi-criteria decision making (FMCDM). For instance, amongst the most multifarious MCDM procedures, the Technique for Order Preference by Similarity to Ideal Solution (TOPSIS) is added to fuzzy logic conditions, explicitly termed as Fuzzy TOPSIS. This method has been flourishingly exercised in various practical and realistic challenging areas like health care, business and manufacturing etc. This chapter highlights the utilization potential of Fuzzy embedded TOPSIS approach in clarifying different MCDM-based obstacles correlated with a real time manufacturing system and other relevant sectors as well. A case study consisting of experimental data sets has been described for better understanding of the suggested methodology. Outcomes of the investigation revealed that, the hybridization of TOPSIS and Fuzzy set theory helped remarkably in realizing the preeminent solution.

Keywords Fuzzy logic · Fuzzy numbers · Multi-criteria decision making · Fuzzy-TOPSIS

A. S. Bhaskar · A. Khan (✉)
Department of Mechanical Engineering, Indian Institute of Information Technology, Design and Manufacturing, Kurnool, Andhra Pradesh 518002, India
e-mail: akhtarkhan00786@gmail.com

A. S. Bhaskar
e-mail: sivabhaskar02@gmail.com

S. R. Patre
Department of Electronics and Communication Engineering, Indian Institute of Information Technology, Design and Manufacturing, Kurnool, Andhra Pradesh 518002, India
e-mail: siturani919@gmail.com

© Springer Nature Switzerland AG 2021
S. Pathak (ed.), *Intelligent Manufacturing*, Materials Forming, Machining and Tribology, https://doi.org/10.1007/978-3-030-50312-3_5

5.1 Introduction

A Multi-Criteria Decision Making (MCDM) which is also acknowledged as Multi-Criteria Decision Analysis (MCDA). In general, an MCMD technique deals with decisions involving the opinion of a best possible solution from several probable prospects in a determination, subject to a few criteria or attribute that may be definite or hazy. Decision-makers (DMs), while making decisions, always try to identify the optimal solution. Lamentably, an optimal solution obtains only in case of single criteria, however absolute decision making involves some contrast or divergence situations. On the other hand, decision making is not just about selecting the best alternative out of the available alternatives. It is often the need to figure all the alternatives for resource allotment and to combine the vigor of the selection of individuals to form a unified selection. In such situations, the concepts of fuzzy logic or fuzzy set theory perform a key role. Fuzzy logic manages through the data originating from computational approach and intelligence. It insists on the involvement of ambiguous human considerations in computing challenges, additionally provides a point for contest decision of multiple criteria and improved valuation of judgments. The most importantly, it is very much convenient to evaluate, judge or come to a decision about the finest interpretation. Altogether, commonly takes advantage of natural language, which doesn't have a definite meaning. Hence, the linguistic are stated in terms of fuzzy numbers, to consider the intuitive opinion of a decision maker in a quantified manner. Most commonly used fuzzy numbers (FN) are triangular FN, trapezoidal FN and Gaussian FN [1]. Latest computation approaches depend on fuzzy logic that allowed evolution of intelligent facilities for decision making, classification, pattern understanding, optimization, and determination. Fuzzy logic is immensely supportive for many people engaged in scientific research and technological development in association with technologists, mathematicians, computer software creators, medical investigators, social savants, and business analysts. Furthermore, it has been exploited in diversified programs like facial paradigm realization, climatiseurs, washing machines, extractors, anti-skid braking systems, conveyance systems, control of underpass systems and knowledge-based controlling networks engineering, graphics processing, commercial automation, artificial intelligence, user electronics, and multi-objective optimizations [2–5].

Furthermore, solving real time manufacturing problems, for instance: selection of a cutting tool material for machining, identifying the best suitable machining parameters to obtain desired machining outputs (viz. decent surface finish, high material removal rate, superior tool life etc.), energy efficiency in automobiles, product design and selection need decisions to take. A decision activity can be planned in three steps: firstly, problem recognition, which involves the identification of the cause of the decision and understanding of the problem to be resolved. Secondly, development of traditional models according to decision maker's preference. Lastly, the formation of action plans as the analysis does not resolve the decision. The purpose of a decision task is to induce information successfully on the decision problem from accessible data, to generate solutions constructively and to furnish a good apprehension about the pattern of a decision problem. The decision maker (DM) has to decide a best alternative from many potential possibilities. However, the decision is based on various criteria or attributes which may be clear or hazy. Moreover, decision making does not depend only on single criteria. On the other hand, soft computing cases like optimization, selection, forecasting, prediction, identification and economics involves multiple criteria, and therefore the individual set of alternative is incapable of providing the best solution. In addition to that, majority of the real time decisions depends on multi criteria. For this reason, Multi-criteria decision making (MCDM) has become so much popular. Usually, decision makers (DMs) while making decisions, inevitably try to specify the optimal solution. Regrettably an optimal solution exists only in case of a single criterion, in actual decision situations nearly any decision includes some discord or disapproval.

A MCDM problem generally contains five modules which are: objective, decision maker's proclivity, options, conventions and consequences respectively. MCDM are invariably compounded due to association of elements involving both engineering and managerial level of analysis. In addition, the MCDM method remains debatable as objectives can lead to unlike solutions. Many methods are used to help people to make their decision according to preferences but the methods themselves cannot make the final decisions. Yoon and Hwang [6], in their study reported a brief outline on multiple attribute decision making (MADM) and various means to model and to solve those problems. Practical real world examples were taken in order to acquaint the readers to expose numerous models for decision making. The authors investigated about various MADM procedures those can be used to address and to solve the multiple decision based problems. Özgen et al. [7] considered a case study for selection of machine tool which involved multiple decision. They proposed a methodology by combining different approaches in MCDM methods like DELPHI, Analytical Hierarchy Process (AHP) and Preference Ranking Organization Method for Enrichment Evaluations (PROMETHEE) with fuzzy set theory. This has given a new opportunity to utilize the hybridization of different MCDM methodologies to real life decision making problems to ensure the diminution of ambiguity in the judgments made by the decision maker. The procedure was applied to select an appropriate machine tool for pressing operation at Turkey, Istanbul. Velasquez M and Hester PT [8], presented a review on MCDM methods and inspected their strengths and weaknesses to propose a clear guide to principles of those methods that can be used in distinctive situations. Palczewski and Sałabun [9], suggested the use of Fuzzy TOPSIS technique to solve various MCDA problems. The current research portrayed a simple and brief review of fuzzy TOPSIS applications. Around 25 different case studies were reported, allied with distinct research areas like environment, business, and supply chain etc. in between 2009 and 2018. In conclusion, cognizance into current trends, most admired approaches, and orientation of study regarding the fuzzy TOPSIS approach were presented. Gandhi and Muruganantham [10], estimated and engaged the potentiality of Multi-criteria Decision Making (MCDM) applications in social media and compared the results. The proposed MCDM based approach effectiveness was examined using a Face book data and the results were compared with existing algorithms like degree, Page Rank and centrality measures. Ultimately, TOPSIS method outperforms all other methods. Tripathy and Tripathy [11], were investigated PMEDM (Powder Mixed Electro-Discharge Machining, an unconventional machining process) to identify the optimal machining parameters. The performance characteristics of H_{11} die steel were determined by adopting TOPSIS method and the results were evaluated using Grey Relations Analysis. It was evident that TOPSIS method can be successfully used for optimization problems in manufacturing systems and further analyzed with ANOVA upon 95% degree of confidence to know the combined implication of the process parameters. Thirumalai and Senthilkumaar [12], addressed a new approach of MCDM perception related to TOPSIS. Their work comprises, high-speed machining of Inconel 718 with carbide cutting tool. The high ranked set of machining parameters were identified as the best set of process parameters in machining Inconel 718 under stated conditions. Singaravel and Selvaraj [13], has taken the Multi objective optimization to a new level by applying the concept to actual industrial problem to ascertain the ideal operational parameters in turning process of EN25 steel with coated carbide tool by make use of embedded Technique for Order Preference by Similarity to Ideal Solution (TOPSIS) and Analytic Hierarchy Process (AHP) method. It is a multiple-objective test for optimization procedure, which was embraced to coexistence of minimizing hardness, surface roughness and maximization of material removal rate (MRR). Consequently, the capability and applicability of this course of action can be extended to major machining processes having many numbers of objectives at a time. Abdel-Kader and Dugdale [14], were demonstrated the conceptual usage and distinguish features of fuzzy logic. In this paper, the authors reported the effective utilization of Triangular fuzzy number with worked examples and their arithmetic operations. Sinha and Sarmah [15], come up with an application of fuzzy interface to study a two stage supply chain management among various supply chain members based on cost and demand criterions. The paper presented a design mechanism to estimate the ambiguity in decision making. The proposed methodology was validated by illustrating a numerical example. Gok [16], was made an attempt to study the effect of cutting parameters in turning ductile iron, considering good surface finish and minimum cutting force as output criteria by taking the advantage of fuzzy embedded with TOPSIS. The results specified, depth of cut has a commanding influence on the surface roughness and cutting force. Yazdani and Payam [17], In this paper, Ashby Procedure and two MCDM methods, VIKOR (Vise Kriterijumska Optimizacija kompromisno Resenja) and TOPSIS were applied to select the appropriate materials for various applications of MEMS. The results have shown a better accordance between the three methods to select material. Finally, comparison between Ashby, VIKOR and TOPSIS methods are presented along with results. Liu [18], studied the application of amalgamation of fuzzy nature with TOPSIS in solving real time challenge in integrated-circuit wafer fabrication process to determine its quality. A fuzzy TOPSIS methodology was developed by taking into consideration the vagueness in quality characteristics to resolve the problem. This proposed methodology can also apply to other industries to investigate different MCDM issues. Nădăban [19], presented a detailed literature review on the general developments of Fuzzy TOPSIS methods along with different application areas. Salih et al. [20], discussed with the state of art of Fuzzy multi-criteria decision making (FMCDM), used in various fields. They researched and presented the taxonomy and gaps in hybridization of fuzzy into MCDM problems. The data consists of literature focusing on FMCDM. Dewangan et al. [21], experimented and examined the dimensional and surface integrity in EDM (Electrical Discharge Machining). The work intended in determining the effect of different EDM variables such as, current pulse, pulse-on time, tool work time and tool lift time influence on the thickness of white layer formed on the surface in accordance with RSM (response surface methodology) and Fuzzy-TOPSIS for identifying the optimal machining parameters. Pavić and Novoselac [22], displayed the simple principle of TOPSIS methodology and its application by considering two case studies examples.

From the literature survey it is clear that, most of the past investigations were mainly focused on highlighting the application potential of various MCDM techniques and hybrid MCDM approached is solving a wide range of problems such as tool selection, material selection, machine selection, robot selection and many more. Some of the collected and mentioned works also indicates the effectiveness and adeptness of various MCDM methods is solving a wide range of problems in almost all the cutting edge areas of the research and academic community. This includes, logistic information technology, energy efficient network selection, analysis of business process outsourcing, airline industries etc. However, the capability of the proposed hybrid FMCDM technique in identifying optimal parametric combination while conventional machining operation is not amply reported so far. Further, very few works have been reported accepting FMCDM methods in explaining machining and related problems specifically turning operations. An attempt has been made in the current study, to amalgamate a popular MCMD approach i.e. TOPSIS with fuzzy logic with an aim to attain the best alternative for maximum production rate in terms of material removal and minimum surface roughness and cutting force during turning operation. The unification of MCDM-based TOPSIS method with fuzzy set theory has been highlighted magnificently. Thus, a hybrid model was established and executed to solve the selected machining problem.

5.2 Fuzzy Set Theory

Fuzzy concept is an extended version of set theory, set theory is provided with the crisp values. It deals with a theory of means in representing uncertainties or the graded concepts in terms of linguistic variables where indistinct theoretical events can be studied perfectively, extensively and analyzed. It could be even studied as a modeling language, well adequate to the situations in which fuzzy relationship, circumstances, and phenomena occur. The concept of Fuzzy set was proposed in 1965 by Dr. Lotfi Zadeh, Zimmermann [23]. He protracted the notation of bilateral membership to address the preconditions and to adapt variant degrees of memberships in continuous real number interval [0, 1] where 0 represents no membership and 1 indicate full membership.

5.2.1 Crisp Set and Fuzzy Set

The main difference between crisp set and fuzzy set is within the representation of their dissimilar membership function. In fuzzy set philosophy, membership function is also termed as characteristic function and represented as $\mu(x)$, $x \in X$. Here μ is recognized as a fuzzy set of universe X and it contains a value in fuzzy set along with membership function $\mu(x)$, $x \in X$. Whereas, a crisp set consists of a unique membership function, on the contrary however as fuzzy set have infinite number of membership functions for their representation. This portrays the usefulness of the fuzzy set over crisp set to each discipline depended on experimental data. Fuzzy sets can be used in mathematical modeling and in finding analytical solutions to solve wide range of problems.

5.2.2 Fuzzy Number

Fuzzy number exists an ordinary number but its value is uncertain. It represents a real number having fuzzy boundary, as the margin of this period is indeterminate, the subset also be a fuzzy number. Fuzzy number can be normalised and convexed. The normalization condition implies stating that maximum membership value as 1 and the convex condition is a continuous line by α-cut.

For example: A fuzzy number $\mu: A \rightarrow [0, 1]$, is a fuzzy set μ of the real line A such that:

1. μ is normal [i.e. there exist $x \in A$ with $\mu(x) = 1$]
2. μ is convex [$(tx + (1 - t)y) \geq \min \{\mu(x), \mu(y)\}$; $x, y \in A$, $0 \leq t \leq 1$].

5.2.2.1 Types of Fuzzy Numbers

The following are majorly used fuzzy numbers:

a. **Trapezoidal fuzzy numbers (TrFN):** If $A = (a_1, a_2, a_3, a_4)$ is a Trapezoidal fuzzy number, it's the characteristic function is given below (Fig. 5.1):

$$\mu(x) = \begin{cases} \frac{x - a_1}{a_2 - a_1}; & a_1 \leq x \leq a_2 \\ 1; & a_2 \leq x \leq a_3 \\ \frac{a_1 - x}{a_4 - a_3}; & a_3 \leq x \leq a_4 \\ 0; & \text{otherwise} \end{cases} \qquad (5.1)$$

Fig. 5.1 Trapezoidal fuzzy
number

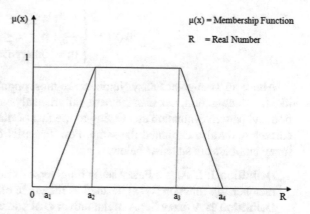

b. **Gaussian fuzzy numbers (GFN)**: Membership function of Gaussian fuzzy numbers is given as below (Fig. 5.2).

$$\mu(x) = e^{\left[-\frac{1}{2}\left(\frac{x-c}{\sigma}\right)^2\right]}; \quad \sigma \in [\sigma_1, \sigma_2] \tag{5.2}$$

where, $c =$ Constant mean and $[\sigma, \sigma_1, \sigma_2]$ are Variable standard deviations.

c. **Triangular Fuzzy Number**: The representative membership function for TFN is given below (Fig. 5.3):

Fig. 5.2 Gaussian Fuzzy
number

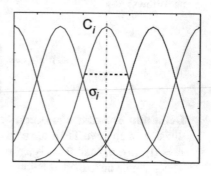

Fig. 5.3 Triangular fuzzy
number

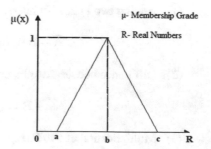

$$\mu(x) = \begin{cases} \frac{x-a}{b-a}; & a \leq x \leq b \\ \frac{c-x}{c-b}; & b \leq x \leq c \\ 0; & \text{otherwise} \end{cases} \qquad (5.3)$$

Above all Triangular Fuzzy Numbers are most popularly applied in various areas like face recognition, risk management, failure analysis, decision-making, optimization and pattern evaluation etc. Owing to the fact of the popularity of the TFNs, the current study also exploited the same. Few essential definitions of fuzzy sets and fuzzy numbers are stated as below:

Definition 1: If $\overline{A}$, is a Fuzzy set in a universe of address X and is labelled by a membership function $\mu_{\overline{A}}(x)$, termed as the grade of membership of x in $\overline{A}$.

Definition 2: A fuzzy Set $\overline{A}$ of the universe of address X is convex if and only if for all x_1, x_2 in X,

$$\mu_{\overline{A}}(\lambda x_1 + (1 - \lambda) x_2) \geq \min(\mu_{\overline{A}}(x_1), \mu_{\overline{A}}(x_2)), \lambda \in [0, 1]$$

Definition 3: A fuzzy set $\overline{A}$ of the universe of address X is called a normal fuzzy set implying that $\exists x_i \in X, \mu_{\overline{A}}(x_i) = 1$.

Definition 4: A fuzzy number is a fuzzy subset in the universe of address X that is both normal and convex.

Definition 5: The triangular fuzzy number (TFN) can be represented as $\overline{A} = (l, m, n)$ and the membership function of the fuzzy number $\overline{A}$ can be characterised as follows:

$$\mu_{\overline{A}}(x) = \begin{cases} 0; & x < l \\ \frac{x-l}{m-l}; & l \leq x \leq m \\ \frac{n-x}{n-m}; & m \leq x \leq n \\ & x > n \end{cases} \qquad (5.4)$$

Definition 6: Arithmetic operations like adding, difference, multiplication and division of any two TFNs are also results into triangular fuzzy number. Consider $\overline{A} = (l_1, m_1, n_1)$ and $\overline{B} = (l_2, m_2, n_2)$, then the arithmetic operations between two TFNs can be described as below:

(1) Adding two Fuzzy Numbers:

$$\overline{A} + \overline{B} = (l_1 + l_2, m_1 + m_2, n_1 + n_2) \qquad (5.5)$$

(2) Difference between two Fuzzy Numbers:

$$\overline{A} - \overline{B} = (l_1 - l_2, m_1 - m_2, n_1 - n_2) \qquad (5.6)$$

(3) Multiplication of two Fuzzy Numbers:

$$\overline{A} * \overline{B} = (l_1 l_2,\ m_1 m_2,\ n_1 n_2) \tag{5.7}$$

(4) Division of two Fuzzy Numbers:

$$\overline{A}/\overline{B} = (l_1/l_2,\ m_1/m_2,\ n_1/n_2) \tag{5.8}$$

Definition 7: Consider $\overline{A} = (l, m, n)$ is a TFN, then the defuzzified value m $(\overline{A})$ can be determined from the following equation:

$$m(\overline{A}) = \frac{l + m + n}{3} \tag{5.9}$$

Definition 8: If $\overline{A} = (l_1, m_1, n_1)$ and $\overline{B} = (l_2, m_2, n_2)$ are two TFNs, then the distance between the two TFNs can be determined from the following equation:

$$d(\overline{A}, \overline{B}) = \sqrt{1/3\big[(l_1 - l_2)^2 + (m_1 - m_2)^2 + (n_1 - n_2)^2\big]} \tag{5.10}$$

5.3 Topsis

The *Technique for Order Preference by Similarity to Ideal Solution* (TOPSIS) was developed by Ching-Lai Hwang and Yoon in 1981 [6]. It is an approach to assess the performance of different choices in the form of resemblance about the best ideal solution. Based on this method, the finest variant is one specifically nearer to the positive-ideal solution but afar from the negative-ideal solution. The positive-ideal solution is served for maximizing the benefit criteria and minimizing the cost criteria whereas the negative-ideal solution chooses for maximizes the cost criteria and minimizes the benefit criteria. In notes, the positive-ideal resolution consists of all foremost values obtainable of criteria, and the negative-ideal resolution provided with all the worst case of values achievable of criteria.

The TOPSIS is one of the classical tools used to solve MCDM-based problems. It was exhaustively implemented to address the MCDM issues associated with distinct areas of manufacturing systems. This modus operandi is favored by many researchers owing to its simplicity and clear understanding. It is made up of two distinct solutions: positive (best) and negative (worst). It contemplates the path to the best (positive ideal) solution and the worst (negative ideal) at the same time.

TOPSIS computation begins with the creation of a decision matrix which contains details of the alternatives and the problem criteria. Decision-makers view the weights of each criterion. Nonetheless, each criterion is ultimately linked to a cost or gain conviction. When solving the problem, the smaller values are preferred when solving the problem for a cost-related criterion, whereas its higher values are preferred for a benefit-related criterion. Next, calculate the standardized decision matrix (NDM) and weighted standard decision matrix (WNDM). The best and worst solutions are computed in the next step. These two solutions are used to evaluate the distance of separation for each criterion. Finally, the relative closeness value of each criterion is determined, and the rank is allocated. The step by step TOPSIS approach is explained as below.

Step 1: Development of decision matrix (DM), by considering m alternatives (experiment runs) and n criterion (experimental outcomes).

$$
\text{Decision Matrix } [X] = \begin{bmatrix}
x_{11} & x_{12} & x_{13} & \cdots & \cdots & x_{1j} & x_{1n} \\
x_{21} & x_{22} & x_{23} & \cdots & \cdots & x_{2j} & x_{2n} \\
x_{31} & x_{32} & x_{33} & \cdots & \cdots & x_{3j} & x_{3n} \\
\vdots & \vdots & \vdots & \vdots & \vdots & \vdots & \vdots \\
x_{i1} & x_{i2} & x_{i3} & \cdots & \cdots & x_{ij} & x_{in} \\
\vdots & \vdots & \vdots & \vdots & \vdots & \vdots & \vdots \\
x_{m1} & x_{m2} & x_{m3} & \cdots & \cdots & x_{mj} & x_{mn}
\end{bmatrix} \tag{5.11}
$$

Here, rows represent the alternatives (i = 1, 2, 3...m) and columns represents the criterion for all corresponding alternatives (j = 1, 2, 3.....n).

Step 2: In order to develop the Normalized Decision Matrix (NDM), each element of the corresponding DM is normalized using the following Eq. (5.12), framing all the resulting values will generate the required NDM and it is represented as X' in Eq. (5.13).

$$
x'_{ij} = \frac{x_{ij}}{\sqrt{\sum_{i=1}^{m} x_{ij}^2}} \tag{5.12}
$$

$$
\text{Normalized Decision Matrix } [X'] = \begin{bmatrix}
x'_{11} & x'_{12} & x'_{13} & \cdots & \cdots & x'_{1j} & x'_{1n} \\
x'_{21} & x'_{22} & x'_{23} & \cdots & \cdots & x'_{2j} & x'_{2n} \\
x'_{31} & x'_{32} & x'_{33} & \cdots & \cdots & x'_{3j} & x'_{3n} \\
\vdots & \vdots & \vdots & \vdots & \vdots & \vdots & \vdots \\
x'_{i1} & x'_{i2} & x'_{i3} & \cdots & \cdots & x'_{ij} & x'_{in} \\
\vdots & \vdots & \vdots & \vdots & \vdots & \vdots & \vdots \\
x'_{m1} & x'_{m2} & x'_{m3} & \cdots & \cdots & x'_{mj} & x'_{mn}
\end{bmatrix}
$$
$$\tag{5.13}$$

Step 3: Weighted Normalized Decision Matrix (WNDM) can be obtained by multiplying the weights of each criterion with the elements of NDM, and it is denoted by Y.

$$
Y = w_j x'_{ij} \tag{5.14}
$$

$$\text{Weighted Normalized Decision Matrix } [Y] = \begin{bmatrix} y_{11} & y_{12} & y_{13} & \cdots & \cdots & y_{1j} & y_{1n} \\ y_{21} & y_{22} & y_{23} & \cdots & \cdots & y_{2j} & y_{2n} \\ y_{31} & y_{32} & y_{33} & \cdots & \cdots & y_{3j} & y_{3n} \\ \vdots & \vdots & \vdots & \vdots & \vdots & \vdots & \vdots \\ y_{i1} & y_{i2} & y_{i3} & \cdots & \cdots & y_{ij} & y_{in} \\ \vdots & \vdots & \vdots & \vdots & \vdots & \vdots & \vdots \\ y_{m1} & y_{m2} & y_{m3} & \cdots & \cdots & y_{mj} & y_{mn} \end{bmatrix}$$

$$(5.15)$$

Step 4: The positive ideal (best) and negative ideal (worst) solutions can be calculated using the following equations.

- Positive ideal solution:

$$A^B = \left\{ (\max y_{ij} \forall j \in J); (\min y_{ij} \forall j \in J') \right\} = \left\{ y_1^B, y_2^B, \ldots y_j^B, \ldots y_n^B, \right\} \quad (5.16)$$

- Negative ideal solution:

$$A^W = \left\{ (\min y_{ij} \forall j \in J); (\max y_{ij} \forall j \in J') \right\} = \left\{ y_1^W, y_2^W, \ldots y_j^W, \ldots y_n^W, \right\}$$

$$(5.17)$$

Where J is the set of best attributes and J' is the set of worst attributes.

Step 5: Each alternative separation measure can be determined by below mentioned formula:

- Separation from positive solution:

$$S_i^B = \sqrt{\sum_{j=1}^{n} \left(y_{ij} - y_j^B \right)^2}; \quad i = 1, 2, 3 \ldots m \quad (18)$$

- Separation from negative solution:

$$S_i^W = \sqrt{\sum_{j=1}^{n} \left(y_{ij} - y_j^W \right)^2}; \quad i = 1, 2, 3 \ldots . m \quad (5.19)$$

Step 6: The relative closeness value to the ideal solution can be obtained from the given formulae:

$$C_i^+ = \frac{S_i^W}{S_i^B + S_i^W}; \quad i = 1, 2, 3 \ldots m; \ 0 \le C_i^+ \le 1 \quad (5.20)$$

Step 7: Ranking can be allocated to all closeness values in descending order. The alternative having larger closeness value is considered as the best performance.

5.4 Fuzzy–TOPSIS

In the following section, a step by step procedure to Fuzzy–TOPSIS (FTOPSIS) approach is proposed. The suggested method utilizes the concepts of fuzzy environment embedded with TOPSIS to manifest the potentiality of the approach in answering multiple criteria decision problems. TOPSIS was enhanced by Chen CT [24] to fuzzy environment, which utilizes a fuzzy linguistic variables as an alternative to the given crisp values in the categorized evaluation. The reformed TOPSIS is a viable method that is equivalent to human thinking in definite environs. The FTOPSIS technique set forth proportionately triumphant exercise samples, particularly in real time problems likes distinctive point of views and convictions are stated by linguistic data, it is recognized to be a better practice than the conventional TOPSIS approach, as it is furnishes a possibility for professionals to convey their reflections in specific periods, so they can evaluate those views without transfiguring into quantified data. Tables 5.1 and 5.2 denote about the various linguistic value assigned to criterions and alternatives respectively.

The present method makes use of linguistics information to communicate the choices of a decision maker. These linguistic detail sets can be additionally utilized

Table 5.1 Linguistic variables for each criterion

Linguistic variable	Notation	Triangular fuzzy number (TFN)
Very very Low	VVL	(0, 0, 0.1)
Very low	VL	(0, 0.1, 0.3)
Low	L	(0.1, 0.3, 0.5)
Medium	M	(0.3, 0.5, 0.7)
High	H	(0.5, 0.7, 0.9)
Very high	VH	(0.7, 0.9, 1)
Very very high	VVH	(0.9, 1, 1)

Table 5.2 Assigned Linguistic Variables for each alterative

Linguistic variable	Notation	Triangular Fuzzy number (TFN)
Very very poor	VVP	(0, 0, 1)
Very poor	VP	(0, 1, 3)
Poor	P	(1, 3, 5)
Fair	F	(3, 5, 7)
Good	G	(5, 7, 9)
Very good	VG	(7, 9, 10)
Very very good	VVG	(9, 10, 10)

for the development of fuzzy decision matrix (FDM) followed by normalized fuzzy decision matrix (NFDM). During the next phase, weighted normalized fuzzy decision matrix is derived, and then fuzzy positive and negative ideal solutions can be decided by acquiring an appropriate weightage for each output criteria. In the end, the distances of each alternative from the fuzzy positive and fuzzy negative ideal solution can be calculated and the preference order of the alternatives can be acquired. Following is the step wise procedure that summarizes FTOPSIS methodology.

Step 1: Organization of a Fuzzy Decision Matrix wherein inputs are expressed in decision matrix as given below. Rows in FDM are alternatives (i = 1, 2, 3... m) and columns denote criterion (j = 1, 2, 3....n).

$$\text{Fuzzy Decision Matrix } [\overline{D}] = \begin{bmatrix} \overline{x}_{11} & \overline{x}_{12} & \overline{x}_{13} & \cdots & \cdots & \overline{x}_{1j} & \overline{x}_{1n} \\ \overline{x}_{12} & \overline{x}_{22} & \overline{x}_{23} & \cdots & \cdots & \overline{x}_{2j} & \overline{x}_{2n} \\ \overline{x}_{31} & \overline{x}_{32} & \overline{x}_{33} & \cdots & \cdots & \overline{x}_{3j} & \overline{x}_{3n} \\ \vdots & \vdots & \vdots & \vdots & \vdots & \vdots & \vdots \\ \overline{x}_{i1} & \overline{x}_{i2} & \overline{x}_{i3} & \cdots & \cdots & \overline{x}_{ij} & \overline{x}_{in} \\ \vdots & \vdots & \vdots & \vdots & \vdots & \vdots & \vdots \\ \overline{x}_{m1} & \overline{x}_{m2} & \overline{x}_{m3} & \cdots & \cdots & \overline{x}_{mj} & \overline{x}_{mn} \end{bmatrix} \quad (5.21)$$

$$W = [W_1 W_2 W_3 W_4]$$

where, W is the weights of the criteria.

Step 2: Formation of Normalized Fuzzy Decision Matrix

$$\text{Normalized Fuzzy Decision Matrix, } [\overline{D'}] = \begin{bmatrix} \overline{r}_{11} & \overline{r}_{12} & \overline{r}_{13} & \cdots & \cdots & \overline{r}_{1j} & \overline{r}_{1n} \\ \overline{r}_{12} & \overline{r}_{22} & \overline{r}_{23} & \cdots & \cdots & \overline{r}_{2j} & \overline{r}_{2n} \\ \overline{r}_{31} & \overline{r}_{32} & \overline{r}_{33} & \cdots & \cdots & \overline{r}_{3j} & \overline{r}_{3n} \\ \vdots & \vdots & \vdots & \vdots & \vdots & \vdots & \vdots \\ \overline{r}_{i1} & \overline{r}_{i2} & \overline{r}_{i3} & \cdots & \cdots & \overline{r}_{ij} & \overline{r}_{in} \\ \vdots & \vdots & \vdots & \vdots & \vdots & \vdots & \vdots \\ \overline{r}_{m1} & \overline{r}_{m2} & \overline{r}_{m3} & \cdots & \cdots & \overline{r}_{mj} & \overline{r}_{mn} \end{bmatrix}$$

$$(5.22)$$

For beneficial criteria and non-beneficial criteria the values of NFDM elements are

$$\overline{r}_{ij} = \left(\frac{l_{ij}}{u_j^+}, \frac{m_{ij}}{u_j^+}, \frac{n_{ij}}{u_j^+} \right) \quad \text{Where, } u_j^+ = \max_i u_{ij} \quad (5.23)$$

$$\overline{r}_{ij} = \left(\frac{l_j^-}{u_{ij}}, \frac{l_j^-}{m_{ij}}, \frac{l_j^-}{l_{ij}} \right) \quad \text{Where, } l_j^- = \min_i l_{ij} \quad (5.24)$$

Step 3: Determine the Weighted Normalized Fuzzy Decision Matrix by multiplying the weights of criteria with the corresponding fuzzy values.

$$\overline{W} = \begin{bmatrix} \overline{w}_{11} & \overline{w}_{12} & \overline{w}_{13} & \cdots & \cdots & \overline{w}_{1j} & \overline{w}_{1n} \\ \overline{w}_{12} & \overline{w}_{22} & \overline{w}_{23} & \cdots & \cdots & \overline{w}_{2j} & \overline{w}_{2n} \\ \overline{w}_{31} & \overline{w}_{32} & \overline{w}_{33} & \cdots & \cdots & \overline{w}_{3j} & \overline{w}_{3n} \\ \vdots & \vdots & \vdots & \vdots & \vdots & \vdots & \vdots \\ \overline{w}_{i1} & \overline{w}_{i2} & \overline{w}_{i3} & \cdots & \cdots & \overline{w}_{ij} & \overline{w}_{in} \\ \vdots & \vdots & \vdots & \vdots & \vdots & \vdots & \vdots \\ \overline{w}_{m1} & \overline{w}_{m2} & \overline{w}_{m3} & \cdots & \cdots & \overline{w}_{mj} & \overline{w}_{mn} \end{bmatrix} \tag{5.25}$$

where:

$$\overline{w}_{ij} = \overline{r}_{ij} * W_j$$

Step 4: Identify the best fuzzy positive ideal solution and worst fuzzy negative ideal solution from the following equations.

$$A^B = \left\{ (max y_{ij} \forall j \in J); (min y_{ij} \forall j \in J') \right\} = \left\{ \overline{w}_1^+, \overline{w}_2^+, \ldots \overline{w}_n^+ \right\} \tag{5.26}$$

$$A^W = \left\{ (min y_{ij} \forall j \in J); (max y_{ij} \forall j \in J') \right\} = \left\{ \overline{w}_1^-, \overline{w}_2^-, \ldots w_n^- \right\} \tag{5.27}$$

where, $\overline{w}_j^+ = (1, 1, 1), \overline{w}_j^- = (0, 0, 0); j = 1, 2, 3, \ldots n.$
J is the set of positive attributes and J' is the set of negative attributes.

Step 5: Compute the distances of each alternative to fuzzy positive and fuzzy negative solutions by using following equations.

$$S_i^B = \sum_{j=1}^{n} d\left(w_{ij} - w_1^+\right); \ i = 1, 2, 3 \ldots m \tag{5.28}$$

$$S_i^W = \sum_{j=1}^{n} d\left(w_{ij} - w_1^-\right); \ i = 1, 2, 3 \ldots m \tag{5.29}$$

where, d is the distance between two fuzzy numbers (refer the Eq. 5.10).

Step 6: Determine the relative closeness value to the ideal solution using the following equation.

$$C_i^+ = \frac{S_i^W}{S_i^B + S_i^W}; \ i = 1, 2, 3 \ldots m \ \ 0 \le C_i^+ \le 1 \tag{5.30}$$

5.5 Case Study

For the better expertise and understanding of the submitted FTOPSIS methodology, a case example has been adopted and described step wise [25]. In the adopted research article, the authors have turned a bar (Commercially Pure Titanium Grade II) diametrically 50 mm and 500 mm long with square shaped uncoated carbide inserts in dry medium. A total of 27 machining trails were performed over a fixed length of 300 mm based on Taguchi's L_{27} orthogonal array design having three factors and three levels were considered for machining. Surface roughness (Ra), material removal rate (MRR) and cutting force (FC) were the required output responses to be measured. Table 5.3 displays data inputted in relation to machining parameters and their levels. While provisional findings were listed in Table 5.4.

The main objective in this study was to determine the best possible machining parameters which can be able to produce excellent surface finish, significant material removal rate and lower cutting force for the given combination of workpiece and cutting tool material. In order to decide the optimal machining parameters, as the problem was based on multi criteria decision making, it might be solved by various MCDM methods. However present proposed method is to expose the potential application zone of Fuzzy TOPSIS methodology within context of manufacturing. The step by step procedure of FTOPSIS was discussed earlier. Accordingly, the first step is to form fuzzy decision matrix from the response table and assigning of weights to the criterion. To assign the weights to the alternatives and criterion, syntactical (linguistic) variables and Triangular fuzzy numbers (TFNs) were utilized. The following tables represent the linguistics assignment to the alternatives and criterion, which depends on the decision maker (DM) and later these linguistics are used to convert the crisp values of DM into fuzzy decision matrix. Tables 5.5 and 5.6 represents the decision matrix having crisp values and fuzzy decision matrix indicating fuzzy linguistic variables respectively.

Now this linguistics has to be converted into corresponding fuzzy numbers. The converted FDM is shown in Table 5.7.

In the forthcoming phase, normalized fuzzy decision matrix is obtained using Eq. 5.22. Table 5.8 portrays the results of normalization.

In the next step, weighted normalized fuzzy decision matrix was evaluated by using Eq. 5.25. Table 5.9 exhibits the weighted normalized fuzzy decision matrix of the current investigation.

Table 5.3 Machining parameters with their corresponding levels

Parameters	Unit	Symbol	Level		
			Low	Medium	High
Cutting speed, v	m/min	v	40	70	110
Feed rate, f	mm/rev	f	0.08	0.1	0.15
Depth of cut, d	mm	d	0.2	0.3	0.4

Table 5.4 Response table of Ra, MRR and F_C

RUN	Input Parameters			Responses		
	V_C	F	D	Ra	MRR	F_C
1	40	0.08	0.2	0.141	0.05	0.197
2	40	0.08	0.3	0.123	0.075	0.202
3	40	0.08	0.4	0.113	0.099	0.233
4	40	0.1	0.4	0.192	0.124	0.242
5	40	0.1	0.2	0.211	0.062	0.216
6	40	0.1	0.3	0.173	0.093	0.218
7	40	0.12	0.3	0.207	0.112	0.232
8	40	0.12	0.4	0.211	0.149	0.244
9	40	0.12	0.2	0.238	0.075	0.206
10	70	0.08	0.3	0.132	0.13	0.19
11	70	0.08	0.4	0.138	0.174	0.208
12	70	0.08	0.2	0.142	0.087	0.167
13	70	0.1	0.2	0.191	0.109	0.183
14	70	0.1	0.3	0.203	0.163	0.204
15	70	0.1	0.4	0.195	0.217	0.209
16	70	0.12	0.4	0.214	0.26	0.2
17	70	0.12	0.2	0.259	0.13	0.189
18	70	0.12	0.3	0.207	0.195	0.191
19	110	0.08	0.4	0.132	0.273	0.148
20	110	0.08	0.2	0.189	0.137	0.135
21	110	0.08	0.3	0.131	0.205	0.146
22	110	0.1	0.3	0.193	0.256	0.162
23	110	0.1	0.4	0.19	0.341	0.17
24	110	0.1	0.2	0.199	0.171	0.136
25	110	0.12	0.2	0.271	0.205	0.147
26	110	0.12	0.3	0.245	0.307	0.173
27	110	0.12	0.4	0.229	0.41	0.182

Finally, the separation measure from each alternative was calculated using Eqs. 5.28 and 5.29. In addition to that, the closeness value was evaluated using Eq. 5.30 corresponding to each alternative. Table 5.10 depicts the separation measure and closeness values. Further, ranking has been assigned by arranging the closeness values in ascending order. In this way, trial number 19 was having the highest closeness value, and hence ranked one.

Table 5.5 Decision matrix (crisp values)

RUN	Ra	MRR	F_C
1	0.141	0.05	0.197
2	0.123	0.075	0.202
3	0.113	0.099	0.233
4	0.192	0.124	0.242
5	0.211	0.062	0.216
6	0.173	0.093	0.218
7	0.207	0.112	0.232
8	0.211	0.149	0.244
9	0.238	0.075	0.206
10	0.132	0.13	0.19
11	0.138	0.174	0.208
12	0.142	0.087	0.167
13	0.191	0.109	0.183
14	0.203	0.163	0.204
15	0.195	0.217	0.209
16	0.214	0.26	0.2
17	0.259	0.13	0.189
18	0.207	0.195	0.191
19	0.132	0.273	0.148
20	0.189	0.137	0.135
21	0.131	0.205	0.146
22	0.193	0.256	0.162
23	0.19	0.341	0.17
24	0.199	0.171	0.136
25	0.271	0.205	0.147
26	0.245	0.307	0.173
27	0.229	0.41	0.182

5.6 Conclusions

In the current chapter, a systematic and effectual amalgamation of fuzzy concepts with TOPSIS methodology (FTOPSIS) has been extrapolated and examined. The capability of the projected methodology been measured in accordance with solving a practical study in identifying optimal machining parameters in order to attain better machinability. Thereby, in the above exploration, a latest hybrid course of action, viz. fuzzy interface combined with TOPSIS has been exploited successfully in both quantitative and qualitative terms of manufacturing criteria. The subsequent concluding observations can be drawn after successful completion of the present investigation:

Table 5.6 Fuzzy decision
Matrix (Linguistic variables)

RUN	Ra	MRR	F_C
1	VG	VVP	F
2	VVG	VVP	P
3	VVG	VVP	VVP
4	F	VVP	VVP
5	P	VVP	VP
6	G	VVP	VP
7	P	VVP	VVP
8	P	VVP	VVP
9	VP	VVP	P
10	VVG	VVP	F
11	VG	VP	P
12	VG	VVP	G
13	F	VVP	F
14	F	VP	P
15	F	P	P
16	P	F	P
17	VVP	VP	F
18	P	VP	F
19	VVG	F	VVG
20	F	VP	VVG
21	VVG	VP	VVG
22	F	P	VG
23	F	G	G
24	F	VP	VVG
25	VVP	P	VVG
26	VP	F	G
27	VP	VVG	G

1. The experimental trail number 19 has found to be the best combination of machining parameters to achieve minimal surface roughness (R_a), maximize material removal rate (MRR) and lowest cutting force(F_C), was perceptible at cutting speed 110 m/min, feed rate 0.08 mm/rev and depth of cut 0.4 mm.
2. Lower surface roughness, low cutting force and high material removal rate were observed at high cutting speed, low feed rate and high depth of cut while turning CP-Ti grade 2 with uncoated carbide insert under dry machining condition.
3. The integration of fuzzy-TOPSIS, utilizing the conceptions of fuzzy set idea, was recognized and appears to be an appropriate and justifiable attempt in accomplishing the optimal combination of machining parameters to meet the industry demand like high production rate without endangering the quality of machining.

Table 5.7 Fuzzy Decision matrix (triangular fuzzy numbers)

RUN	Ra	MRR	F_C
1	(7, 9, 10)	(0, 0, 1)	(3, 5, 7)
2	(9, 10, 10)	(0, 0, 1)	(1, 3, 5)
3	(9, 10, 10)	(0, 0, 01)	(0, 0, 1)
4	(3, 5, 7)	(0, 0, 1)	(0, 0, 1)
5	(1, 3, 5)	(0, 0, 1)	(0, 1, 3)
6	(5, 7, 9)	(0, 0, 1)	(0, 1, 3)
7	(1, 3, 5)	(0, 0, 1)	(0, 0, 1)
8	(1, 3, 5)	(0, 0, 1)	(0, 0, 1)
9	(0, 1, 3)	(0, 0, 1)	(1, 3, 5)
10	(9, 10, 10)	(0, 0, 1)	(3, 5, 7)
11	(7, 9, 10)	(0, 1, 3)	(1, 3, 5)
12	(7, 9, 10)	(0, 0, 1)	(5, 7, 9)
13	(3, 5, 7)	(0, 0, 1)	(3, 5, 7)
14	(3, 5, 7)	(0, 1, 3)	(1, 3, 5)
15	(3, 5, 7)	(1, 3, 5)	(1, 3, 5)
16	(1, 3, 5)	(3, 5, 7)	(1, 3, 5)
17	(0, 0, 1)	(0, 1, 3)	(3, 5, 7)
18	(1, 3, 5)	(0, 1, 3)	(3, 5, 7)
19	(9, 10, 10)	(3, 5, 7)	(9, 10, 10)
20	(3, 5, 7)	(0, 1, 3)	(9, 10, 10)
21	(9, 10, 10)	(0, 1, 3)	(9, 10, 10)
22	(3, 5, 7)	(1, 3, 5)	(7, 9, 10)
23	(3, 5, 7)	(5, 7, 9)	(5, 7, 9)
24	(3, 5, 7)	(0, 1, 3)	(9, 10, 10)
25	(0, 0, 1)	(1, 3, 5)	(9, 10, 10)
26	(0, 1, 3)	(3, 5, 7)	(5, 7, 9)
27	(0, 1, 3)	(9, 10, 10)	(5, 7, 9)

Nevertheless, this method might be bounded to the designated range of process variables.

4. The proposed methodology can also be used to work out similar type of problematical issues in real time manufacturing systems.

Table 5.8 Normalized Fuzzy
Decision Matrix

RUN	Ra	MRR	F_C
1	(0.7, 0.9, 1)	(0, 0, 0.1)	(0.3, 0.5, 0.7)
2	(0.9, 1, 1)	(0, 0, 0.1)	(0.1, 0.3, 0.5)
3	(0.9, 1, 1)	(0, 0, 0.1)	(0, 0, 0.1)
4	(0.3, 0.5, 0.7)	(0, 0, 0.1)	(0, 0, 0.1)
5	(0.1, 0.3, 0.5)	(0, 0, 0.1)	(0.1, 0.3, 0.5)
6	(0.5, 0.7, 0.9)	(0, 0, 0.1)	(0.1, 0.3, 0.5)
7	(0.1, 0.3, 0.5)	(0, 0, 0.1)	(0, 0, 0.1)
8	(0.1, 0.3, 0.5)	(0, 0, 0.1)	(0, 0, 0.1)
9	(0.1, 0.3, 0.5)	(0, 0, 0.1)	(0.1, 0.3, 0.5)
10	(0.9, 1, 1)	(0, 0, 0.1)	(0.3, 0.5, 0.7)
11	(0.7, 0.9, 1)	(0, 0.1, 0.3)	(0.1, 0.3, 0.5)
12	(0.7, 0.9, 1)	(0, 0, 0.1)	(0.5, 0.7, 0.9)
13	(0.3, 0.5, 0.7)	(0, 0, 0.1)	(0.3, 0.5, 0.7)
14	(0.3, 0.5, 0.7)	(0, 0.1, 0.3)	(0.1, 0.3, 0.5)
15	(0.3, 0.5, 0.7)	(0.1, 0.3, 0.5)	(0.1, 0.3, 0.5)
16	(0.1, 0.3, 0.5)	(0.3, 0.5, 0.7)	(0.1, 0.3, 0.5)
17	(0, 0, 0.1)	(0.1, 0.3, 0.5)	(0.3, 0.5, 0.7)
18	(0.1, 0.3, 0.5)	(0.1, 0.3, 0.5)	(0.3, 0.5, 0.7)
19	(0.9, 1, 1)	(0.3, 0.5, 0.7)	(0.9, 1, 1)
20	(0.3, 0.5, 0.7)	(0, 0.1, 0.3)	(0.9, 1, 1)
21	(0.9, 1, 1)	(0, 0.1, 0.3)	(0.9, 1, 1)
22	(0.3, 0.5, 0.7)	(0.1, 0.3, 0.5)	(0.7, 0.9, 1)
23	(0.3, 0.5, 0.7)	(0.5, 0.7, 0.9)	(0.5, 0.7, 0.9)
24	(0.3, 0.5, 0.7)	(0, 0.1, 0.3)	(0.9, 1, 1)
25	(0, 0, 0.1)	(0.1, 0.3, 0.5)	(0.9, 1, 1)
26	(0, 0.1, 0.3)	(0.3, 0.5, 0.7)	(0.5, 0.7, 0.9)
27	(0, 0.1, 0.3)	(0.9, 1, 1)	(0.5, 0.7, 0.9)

Table 5.9 Weighted Normalized Fuzzy Decision Matrix

RUN	Ra	MRR	F_C
1	(0.49, 0.81, 1)	(0, 0, 0.1)	(0.21, 0.45, 0.7)
2	(0.63, 0.9, 1)	(0, 0, 0.1)	(0.07, 0.27, 0.5)
3	(0.63, 0.9, 1)	(0, 0, 0.1)	(0, 0, 0.1)
4	(0.21, 0.45, 0.7)	(0, 0, 0.1)	(0, 0, 0.1)
5	(0.07, 0.27, 0.5)	(0, 0, 0.1)	(0.07, 0.27, 0.5)
6	(0.35, 0.63, 0.9)	(0, 0, 0.1)	(0.07, 0.27, 0.5)
7	(0.07, 0.27, 0.5)	(0, 0, 0.1)	(0, 0, 0.1)
8	(0.07, 0.27, 0.5)	(0, 0, 0.1)	(0, 0, 0.1)
9	(0.07, 0.27, 0.5)	(0, 0, 0.1)	(0.07, 0.27, 0.5)
10	(0.63, 0.9, 1)	(0, 0, 0.1)	(0.21, 0.45, 0.7)
11	(0.49, 0.81, 1)	(0, 0.1, 0.3)	(0.07, 0.27, 0.5)
12	(0.49, 0.81, 1)	(0, 0, 0.1)	(0.35, 0.63, 0.9)
13	(0.21, 0.45, 0.7)	(0, 0, 0.1)	(0.21, 0.45, 0.7)
14	(0.21, 0.45, 0.7)	(0, 0.1, 0.3)	(0.07, 0.27, 0.5)
15	(0.21, 0.45, 0.7)	(0.09, 0.3, 0.5)	(0.07, 0.27, 0.5)
16	(0.07, 0.27, 0.5)	(0.27, 0.5, 0.7)	(0.07, 0.27, 0.5)
17	(0, 0, 0.1)	(0.09, 0.3, 0.5)	(0.21, 0.45, 0.7)
18	(0.07, 0.27, 0.5)	(0.09, 0.3, 0.5)	(0.21, 0.45, 0.7)
19	(0.63, 0.9, 1)	(0.27, 0.5, 0.7)	(0.63, 0.9, 1)
20	(0.21, 0.45, 0.7)	(0, 0.1, 0.3)	(0.63, 0.9, 1)
21	(0.63, 0.9, 1)	(0, 0.1, 0.3)	(0.63, 0.9, 1)
22	(0.21, 0.45, 0.7)	(0.09, 0.3, 0.5)	(0.49, 0.81, 1)
23	(0.21, 0.45, 0.7)	(0.45, 0.7, 0.9)	(0.35, 0.63, 0.9)
24	(0.21, 0.45, 0.7)	(0, 0.1, 0.3)	(0.63, 0.9, 1)
25	(0, 0, 0.1)	(0.09, 0.3, 0.5)	(0.63, 0.9, 1)
26	(0, 0.09, 0.3)	(0.27, 0.5, 0.7)	(0.35, 0.63, 0.9)
27	(0, 0.09, 0.3)	(0.81, 1, 1)	(0.35, 0.63, 0.9)

Table 5.10 Overall assessment table

RUN	Speed	Feed	Depth of cut	Closeness	Rank
1	40	0.08	0.2	0.0158	7
2	40	0.08	0.3	0.0150	9
3	40	0.08	0.4	0.0165	3
4	40	0.1	0.4	0.0140	10
5	40	0.1	0.2	0.0106	21
6	40	0.1	0.3	0.0134	12
7	40	0.12	0.3	0.0126	15
8	40	0.12	0.4	0.0126	15
9	40	0.12	0.2	0.0106	21
10	70	0.08	0.3	0.0164	4
11	70	0.08	0.4	0.0112	19
12	70	0.08	0.2	0.0173	2
13	70	0.1	0.2	0.0133	13
14	70	0.1	0.3	0.0086	26
15	70	0.1	0.4	0.0100	24
16	70	0.12	0.4	0.0101	23
17	70	0.12	0.2	0.0076	27
18	70	0.12	0.3	0.0100	24
19	110	0.08	0.4	0.0189	1
20	110	0.08	0.2	0.0131	14
21	110	0.08	0.3	0.0161	5
22	110	0.1	0.3	0.0139	11
23	110	0.1	0.4	0.0159	6
24	110	0.1	0.2	0.0126	17
25	110	0.12	0.2	0.0107	20
26	110	0.12	0.3	0.0116	18
27	110	0.12	0.4	0.0152	8

References

1. D. Dubois, H. Prade, Fuzzy numbers: an overview, in *Readings in Fuzzy Sets for Intelligent Systems* (Elsevier, 1993), pp. 112–148
2. K.-M. Lim, Y.-C. Sim, K.-W. Oh, *A face recognition system using fuzzy logic and artificial neural network*, in *[1992 Proceedings] IEEE International Conference on Fuzzy Systems* (IEEE, 1992)
3. P. Melin et al., Edge-detection method for image processing based on generalized type-2 fuzzy logic. IEEE Trans. Fuzzy Syst. **22**(6), 1515–1525 (2014)
4. C. von Altrock, J. Gebhardt, *Recent successful fuzzy logic applications in industrial automation*, in *Proceedings of IEEE 5th International Fuzzy Systems* (IEEE, 1996)

5. A. Saffiotti, The uses of fuzzy logic in autonomous robot navigation. Soft. Comput. **1**(4), 180–197 (1997)
6. K.P. Yoon, C.-L. Hwang, *Multiple Attribute Decision Making: An Introduction*, vol. 104 (Sage, 1995)
7. A. Özgen et al., A multi-criteria decision making approach for machine tool selection problem in a fuzzy environment. Int. J. Comput. Intell. Syst. **4**(4), 431–445 (2011)
8. M. Velasquez, P.T. Hester, An analysis of multi-criteria decision making methods. Int. J. Oper. Res. **10**(2), 56–66 (2013)
9. K. Palczewski, W. Sałabun, The fuzzy TOPSIS applications in the last decade. Procedia Comput. Sci. **159**, 2294–2303 (2019)
10. M. Gandhi, A. Muruganantham, Potential influencers identification using multi-criteria decision making (MCDM) methods. Procedia Comput. Sci. **57**, 1179–1188 (2015)
11. S. Tripathy, D. Tripathy, Multi-attribute optimization of machining process parameters in powder mixed electro-discharge machining using TOPSIS and grey relational analysis. Eng. Sci. Technol. Int. J. **19**(1), 62–70 (2016)
12. R. Thirumalai, J. Senthilkumaar, Multi-criteria decision making in the selection of machining parameters for Inconel 718. J. Mech. Sci. Technol. **27**(4), 1109–1116 (2013)
13. B. Singaravel, T. Selvaraj, Optimization of machining parameters in turning operation using combined TOPSIS and AHP method. Tehnicki Vjesnik **22**(6), 1475–1480 (2015)
14. M.G. Abdel-Kader, D. Dugdale, Evaluating investments in advanced manufacturing technology: a fuzzy set theory approach. Br. Account. Rev. **33**(4), 455–489 (2001)
15. S. Sinha, S. Sarmah, An application of fuzzy set theory for supply chain coordination. Int. J. Manag. Sci. Eng. Manag. **3**(1), 19–32 (2008)
16. A. Gok, A new approach to minimization of the surface roughness and cutting force via fuzzy TOPSIS, multi-objective grey design and RSA. Measurement **70**, 100–109 (2015)
17. M. Yazdani, A.F. Payam, A comparative study on material selection of microelectromechanical systems electrostatic actuators using Ashby, VIKOR and TOPSIS. Mater. Des. (1980–2015), **65**, 328–334 (2015)
18. C.-M. Liu, M.-Y. Ji, W.-C. Chuang, Fuzzy TOPSIS for multiresponse quality problems in wafer fabrication processes. Adv. Fuzzy Syst. **2013** (2013)
19. S. Nădăban, S. Dzitac, I. Dzitac, Fuzzy topsis: a general view. Procedia Comput. Sci. **91**, 823–831 (2016)
20. M.M. Salih et al., Survey on fuzzy TOPSIS state-of-the-art between 2007 and 2017. Comput. Oper. Res. **104**, 207–227 (2019)
21. S. Dewangan, S. Gangopadhyay, C. Biswas, Study of surface integrity and dimensional accuracy in EDM using Fuzzy TOPSIS and sensitivity analysis. Measurement **63**, 364–376 (2015)
22. Z. Pavić, V. Novoselac, Notes on TOPSIS method. Int. J. Res. Eng. Sci. **1**(2), 5–12 (2013)
23. Zimmermann, H.-J., *Fuzzy set theory—and its applications*. 2011: Springer Science & Business Media
24. C.-T. Chen, Extensions of the TOPSIS for group decision-making under fuzzy environment. Fuzzy Sets Syst. **114**(1), 1–9 (2000)
25. A. Khan, K. Maity, A novel MCDM approach for simultaneous optimization of some correlated machining parameters in turning of CP-titanium grade 2. Int. J. Eng. Res. Afr. (2016)

Chapter 6
Multi-criteria Decision Making Through Soft Computing and Evolutionary Techniques

Senol Bayraktar and Kapil Gupta

Abstract Time, cost and quality factors should be taken into consideration to increase productivity in production. Innovative approaches and solutions in manufacturing can be obtained by controlling the independent variables affecting these factors. For this reason, the use of optimization techniques based on different algorithm structures is increasing. Multi-criteria decision-making (MCDM) tools such as ANN (Artificial neural network), FL (Fuzzy logic), GA (Genetic algorithm), PSO (Particle swarm optimization), GRA (Grey relational analyses), TOPSIS (Technique for order of preference by similarity to ideal solution), PROMETHEE (Preference Ranking Organization Method for Enrichment Evaluation), AHP (Analytic Hierarchy Process), ELECTRE (Elimination Et Choix Traduisant la REaite) and hybrid are commonly used. Particularly, it is preferred in comparative analysis in the literature for optimum parameter determination and prediction of results in machinability studies. Throughout this chapter, research based on the studies on multi-criteria decision-making tools is discussed. Moreover, various characteristics and difference among these tools are also reported.

Keywords Decision making · Optimization · ANN · Fuzzy logic · PSO · GRA · TOPSIS · AHP · ELECTRE · PROMETHEE

6.1 Introduction

Multi-Criteria Decision Making (MCDM) is one of the rapidly developing areas in recent years due to changes in the business sector. Decision-makers (DMr) need decision support to determine among choices and often erase less preferred options. Decision making (DM) using computers is widely preferred in all areas of DM. Since

S. Bayraktar (✉)
Faculty of Engineering, Department of Mechanical Engineering, Recep Tayyip Erdogan University, Rize, Turkey
e-mail: senol.bayraktar@erdogan.edu.tr

K. Gupta
Department of Mechanical and Industrial Engineering Technology, University of Johannesburg, Johannesburg, Republic of South Africa

© Springer Nature Switzerland AG 2021
S. Pathak (ed.), *Intelligent Manufacturing*, Materials Forming, Machining and Tribology, https://doi.org/10.1007/978-3-030-50312-3_6

MCDM is preferred in many fields, it has contributed to the emergence of different methods. The MCDM process has become easier for DMr, even in complex mathematical solutions, with the increasing use of computers in recent years [1]. DM is the main factor for success in many points where information and knowledge must be addressed in any working discipline. The processes and procedures consist of tasks and requirements that cover many factors and aspects to consider. It is difficult to decide and difficult to deal with in this case. Therefore, a demand arises for a mechanism that should assist in solving complex scenarios. MCDM has been developed to facilitate solving problems under different conditions and application areas [2–4]. As an example, the decisions taken for the water resources method are usually guided by multi-objective determined in financial and non-financial situations. Outputs generally consist of elements such as biodiversity, recreation, landscape and human health. Therefore, these features make attractive the use of MCDM in the planning of water resources [5]. The MCDM technique is also widely preferred for the aviation industry. The growth in the aviation industry makes a positive contribution in terms of offering different options for passengers. However, air transportation also causes adverse environmental effects.

For this reason, society tries to meet this continually growing traffic and its effects using different opposing criteria. The demand for air transportation is increasing day by day. Accordingly, the airline company must decide which aircraft to use on any route and how to choose the partner. It also has to decide on options such as how to set appropriate service levels and how to improve operational efficiency. Therefore, MCDM method is preferred for the solution of the problems in DM in different categories that may arise in various fields [6].

Manufacturing technologies have gone through continuous self-renewing stages. Product-based, rapidly changing techniques are useful in driving sustainable control mechanisms in the manufacturing industry. In order to overcome various challenges, the manufacturing industry needs to make a good selection of machinery and equipment, as well as appropriate manufacturing strategies, product designs, manufacturing processes, workpiece and tool materials [7]. In modern manufacturing, manufacturers attach great importance to factors such as better product design, product quality, lower cost, quick response to market changes, higher customer satisfaction and shortening the introduction time of new products to increase competition. In terms of the competition at the global level, all activities require systematic and integrated planning and optimization. Practical problems in the production environment can be formulated and solved by multiple-criteria optimization methods. In other words, in light of the data obtained from the experimental results, problems can be solved with mathematical models based on appropriate optimization techniques and algorithms. In addition to Taguchi method, soft computing methods such as ANN [8], response surface methodology (RSM), GA, GRA, TOPSIS, PSO, FL, AHT, PROMETHEE, ELECTRE are used for this purpose.

6.2   Significance of Decision Making

There are different alternatives in every person's life. Therefore, one has to meet the DM need. This necessity of decision reveals that any situation has various options and is related to the selection of one of them. The DM process consists of evaluating alternatives in many cases and choosing the most preferred of them. Making the right decision means that an overall value selected from all factors and conflicting requirements can be optimized. In other words, the target sought as much as possible will be reached at maximum level [9]. DM is gaining importance, especially in industrial sectors, where time and cost factors are at the forefront in terms of increasing the competitiveness of enterprises. In this way, criteria, alternatives and solution methods can be determined accurately to achieve the targets and faster and optimum results can be obtained.

6.2.1   MCDM Mechanism

The required conditions for an effective MCDM are the determination of the factors affecting the production environment, the investigation of the nature of DM, the acquisition of different methods and techniques, and the construction of a DM approach for issues related to design, planning and management of manufacturing systems [7, 10].

$$W = [w_1, w_2 \ldots w_n]$$

According to Table 6.1, $A_1, A_2, \ldots, A_n$, the alternatives that the DMr have to select, $B_1, B_2, \ldots, B_n$ is the criterion where choice efficiency is calculated, x_{ij} is the degree of choice A_i according to Bj, wj is the weight of the Bj criterion. MCDM's process planning is done in the following order:

(a)   Determination of the evaluation criteria that can bring solutions appropriate to the objectives,
(b)   Derivation of alternative systems to achieve the goals,
(c)   Determining of alternatives considering measures,
(d)   Application of MC analysis technique,

Table 6.1   Matrix format of MCDM

	B_1	B_2	…	B_n
A_1	x_{11}	x_{12}	…	x_{1n}
A_2	x_{21}	x_{22}	…	X_{2n}
⋮	⋮	⋮	⋮	⋮
A_n	x_{m1}	x_{m2}	…	X_{mn}

(e) Optimizing an alternative,
(f) If the final solution is not valid, repeat the MC optimization by gathering new information.

Here, a and e steps are processes in which decision-makers play a central role at the maximum, while others are mostly related to engineering issues [1]. Generally, MCDM methods such as TOPSIS, ANN, GA, PSO, GRA, FL, AHT, PROMETHEE and ELECTRE are used to perform these steps.

6.3 Decision Making Approaches and Applications

6.3.1 Artificial Neural Network (ANN)

ANN is commonly preferred in the formation of many estimations and decision models. Computer systems develop information production automatically. It has a structure designed for events that are almost impossible to program. In a multilayer and feed-forward neural network, artificial neural cells connect to form a sequential layered structure. Their task is to collect, store and generalize information. Neural cells are known as process elements. Each processing element has five characteristics. These are input, weights, sum function, activation function and output [11]. Process elements are connected through networks. The elements and the connections among them form the artificial neural network. The weights of the connections are determined in the learning process of the network. Sum function (u) is calculated according to Eq. (6.1). Here, x represents the input values, w denotes the weights and b represents the bias.

$$u = \sum_{i=1}^{n} w_i x_i + b \qquad (6.1)$$

Activation functions in the neural network may have different properties. For example, the sigmoid function is calculated as in Eq. (6.2). Here, the y value between 0 and 1 is considered the output value.

$$y = \frac{1}{1 + e^{-u}} \qquad (6.2)$$

Artificial neural networks include information gathering, training, testing and prediction. In the first stage, the data is obtained. The data collected is then preprocessing. The aim is to make the same things that represent knowledge, similar and different things differently. The connection between the input and output data is formed in the training stage (Fig. 6.1). If the same data is used, the data is considered highly trained. The network will always recognize the same data. This means over-trained data. It also has a negative impact on the effectiveness of the results.

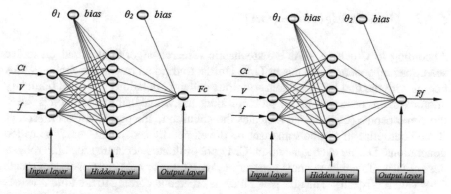

Fig. 6.1 Flow diagram of ANN [13]

Therefore, it is recommended to use different data. Some of the data is also reserved for testing. Thus, the presence of overtraining is determined by the test results. If positive results are obtained during the testing process, it is understood that the connections between the networks are correct. If the results are not accurate, the training of the data continues until the appropriate network structures are found. After the training and testing of the data, the accuracy of the models obtained with different data is determined. Thus, by the previously known data in ANN, the prediction of the intermediate values after training and testing can be calculated according to the determined mathematical models [12].

Case study-1

Das et al. [14] predicted the cutting force using ANN for milling Al-4.5Cu alloy, Al-4.5Cu-TiC and Al-4.5Cu-SiC metal matrix composites with robust carbide tools. They used Log-Sigmoid activation function for ANN, Levenberg-Marquardt for training algorithm and Minimum MSE for loss function criteria. The ANN model for surface roughness (SR) and cutting force (CF) was found to be entirely compatible with the experimental and predicted data with minimum error. They stated that feed-forward backpropagation type ANN could be used to predict CF and SR.

Case study-2

Orak et al. [11] developed ANN-TOPSIS based DM to find the highest stable depth of cut (DOC) in turning. It was stated that the ANN model was developed during the training stage, and the weights attained from the ANN model were used to demolish the negligible variables. In addition, the model has been developed by using TOPSIS with eight variables affecting the depth of cut. It was determined that the experiments with the highest TOPSIS scores in the model had the highest constant cutting depth, and the experimental results were highly compatible with the developed model.

6.3.2   Genetic Algorithm (GA)

According to Goldberg, GAs are stochastic research algorithms based on native selection and genetic mechanisms [15]. Unlike traditional research techniques, GA begins with the first set of random solutions called populations that meet boundary and/or system constraints for the problem. Each individual in the population is called the chromosome (CM), which indicates the solution to the problem. Chromosomes (CMs) comprise of symbols and can be developed by sequential iterations called generations. During each generation, CMs are evaluated according to some robustness criteria. To create the next generation, mutations or crossover operators create CMs called offspring. The new generation is created according to the fitness values of some of the parents and offsprings. In order to keep the size of population constant, others are rejected. Fitter CMs have the highest probability of selection. Thus, after several generations, the algorithms combine best on the CM [16]. GA operations generally consist of selection, adaptive crossover and mutation stages (Fig. 6.2).

Selection

In the selection phase, the goal is to pass optimization to new generations either directly or by cross-matching. Accordingly, the probability of selecting an individual

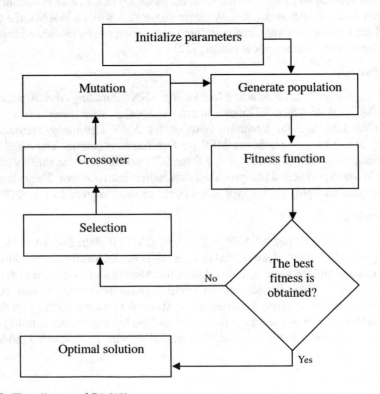

Fig. 6.2 Flow diagram of GA [18]

can be calculated as follows. N is the population size, and F_i is the individual fitness value in Eq. (6.3).

$$P_i = \frac{F_i}{\sum_{k=1}^{N} F_N} \tag{6.3}$$

Adaptive crossover

Parent individuals are randomly chosen by the crossover operator to create offspring individuals (OIs). The number of OIs is assigned by crossover probability PC. Therefore, the self-adaptive penalty function is proposed to develop the research capability of the algorithm. The description of this statement can be written as Eq. (6.4).

$$P_c(t+1) = \begin{cases} P_c(t) x \frac{(F\max - F)}{F\max - F_{avg}}, F \geq F_{avg} \\ P_c(t), F < F_{avg} \end{cases} \tag{6.4}$$

where F_{max} is the maximum fitness value, F_{avg} is the average fitness value in each individual, F is the largest of the two parents' fitness value, and $P_c(t)$. It represents the generation's crossover ratio. Crossover is applied repeatedly until a new population is formed within the system [17].

Mutation

The crossover operator is used to combine existing genes to obtain new CMs, while the mutation operator creates new CMs by causing minor degradation of genes. Therefore, mutation helps to increase population variation to expand the space solution. For each string element in the mating pool, GA controls whether it mutates. Replaces random item value if necessary [17].

Case study-1

Shrivastava and Pandey [19] studied the parametric multi-objective optimization (MOO) of quality characteristics with hybrid, multiple regression and genetic algorithm approach in laser cutting of Inconel 718. The objective function is defined in M-files and MATLAB encoding. Considering different quality characteristics, second order regression models were used as an objective function within the parameter ranges. For the determination of quality characteristics, simultaneous critical optimization parameters based on literature and trial and error method were preferred. For this purpose, the population was selected as size 200, population type double vector, crossover probability 0.8, mutation probability 0.07, maximum generation 800, initial population 200 and initial scores 200. When the experimental results were compared with optimization, it was found that 88%, 10.63% and 42.15% improvement was achieved for kerf deviation, kerf width and kerf taper, respectively.

Case study-2

Mia and Dhar [20] performed the estimation and optimization of surface roughness by GA, SVR (Support vector regression) and RSM methods in hard turning of tempered AISI 1060 steel under influential cooling conditions. The input variables for the model formulation were selected as cutting speed (CS), feed rate (FR) and material hardness (MH). For determination of control factor corresponding to minimum surface roughness, separate optimization models were determined using GA and composite desirability function. It is recommended that both optimization models use a CS of 161 m/min, 0.1 mm/rev FR and MH of ~43 HRC for minimum surface roughness.

6.3.3 Fuzzy Logic (FL)

FL is one of the optimization methods used for the nearest estimation of the obtained numerical data. In practice, the expert can experimentally specify the quality of the outputs according to the input parameters. By using the FL approach, the expert's effort to find specific output results can be eliminated. Much of the knowledge is uncertain and unclear. Therefore, it is a fact that all real systems contain incomplete and inaccurate information. It is appropriate to use an FL method based on fuzzy sets to solve such situations [21]. FL has been successfully applied in the modelling of complex and uncertain problems in science and engineering [22]. An exemplary fuzzy logic method with a single input and a single output (SISO) is an example of a fuzzy algorithm method that defines the generation of a voltage signal supplied from the microcontroller to the transistor, depending on the temperature in the cutting area. The voltage value obtained from the microcontroller is defined according to $U = f(t)$. For input and output variables, the membership function has a triangular shape. Input and output values are determined according to the following equations.

$$\mu_{t_i} = \begin{cases} 0 & if\ t_i < a\ and\ t_i > c, \\ \frac{t_i-a}{b-a}, & if\ a < t_i < b, \\ \frac{c-t_i}{c-b}, & if\ b < t_i < c \end{cases} \tag{6.5}$$

t_i represents the input variable. a, b and c values are selected from the values of the triangular membership function. t_1, t_2, t_3, t_4 and t_5 are variable input terms. Sigmoid membership function can be applied according to Eq. (6.7). Accordingly, the values of a, b and c are the parameters of the shaped-shaped membership function, and x is the quantitative value of the input parameter.

$$t = [t_1] + [t_2] + [t_3] + [t_4] + [t_5] \tag{6.6}$$

$$\mu(y) = \left[1 + \left(\frac{x - c}{a}\right)^b\right]^{-1} \tag{6.7}$$

According to the self-calibrated fuzzy algorithm developed for cutting temperature, operations can be performed in three stages as follows.

- *Fuzzification of input data*

The triangular membership function is performed to define the input and output parameters. The fuzzy control rules (FCR) for the control system are listed as follows. Accordingly, when the system is running, the applied process determines the parameters that affect the accuracy level for each assumption of FCR $\alpha_1, ..., \alpha_5$.

$$FCR1 : \text{IF "}t \text{ is } t_1\text{" THEN "}u \text{ is } U_5\text{"}$$
$$FCR2 : \text{IF "}t \text{ is } t_2\text{" THEN "}u \text{ is } U_4\text{"}$$
$$\vdots$$
$$FCR5 : \text{IF "}t \text{ is } t_5\text{" THEN "}u \text{ is } U_1\text{"}$$

- *Logical derivation*

This stage is performed in two phases. The first is the implication stage, and the other is the complication step. Accordingly, the truncation levels of derivations for each of the five FCRs are determined using soft arithmetic operation and $a_1 \ldots a_5 = \min_{\delta}(t; \Delta t)$ to find the min.

$$\min_{\delta}(t; \Delta t) = \frac{t + \Delta t + \delta^2 - \sqrt{(1 - \Delta t)^2 + \delta^2}}{2} \tag{6.8}$$

For each output, term is determined according to Eq. (6.11).

$$U'u = \bigvee_{i=1}^{n} U'_i = a_i \bigwedge U_i \tag{6.9}$$

$a_i = \max(U'_i)$ is combined with the truncated membership functions in the composition stage.

- *Defuzzification*

The output value of the system is calculated according to Eq. (6.10) in this step. The flow diagram of the system operation of the fuzzy logic method is given in Fig. 6.3.

$$y_{out} = \sum_{i=1}^{n} U'u / \sum_{i=1}^{n} u \tag{6.10}$$

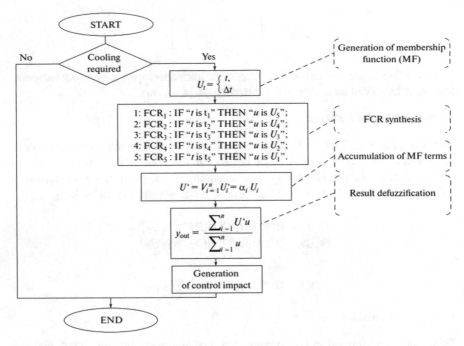

Fig. 6.3 Flow diagram of system operation in the fuzzy logic method [23]

Case Study-1

Hossain et al. [24] developed an FL based estimation model for the estimation of kerf width in laser beam machining. In order to estimate the kerf width in FL, gas pressure, laser power, CS and standoff distance and the fuzzy toolbox Mamdani technique in MATLAB R2009b were used as input parameters. The relative error was 3.852% and $R^2 = 0.989$ in the fuzzy model training for kerf width, while in the test phase, the relative error was 5.2% and $R^2 = 0.966$. According to the FL model, the minimum kerf width was calculated as 0.3167 mm under 0.5 bar gas pressure, 100 W laser power, 1 mm standoff distance and 1.2 m/min cutting speed.

Case Study-2

Kuram and Özçelik [25] the micro-milling efficiency of AISI 304 stainless steel modelled using FL and Taguchi based regression method. The input parameters were CS, FT and DOC, while the output parameters were tool wear, cutting force and surface roughness. In the FL approach, the minimum and maximum values of the input and output parameters were first determined. Fuzzification was then performed to obtain fuzzy sets, and numerical values turned into linguistic expressions such as low, medium and high for CS, FR and DOC. Gaussian membership function was used to fuzzification the input and output variables. According to experimental results, it has been suggested that FL and regression modelling can be used to estimate cutting forces, tool wear and surface roughness.

6.3.4 Particle Swarm Optimization (PSO)

PSO was developed by Kennedy and Eberhart [26, 27] by imitating the hunting behaviour of birds and fish. The bird in the group is assumed as a solution in the research field of the optimization problem and is known as the 'particle'. The objective function is used to compare the strength value of each part. Each particle has its own best solution in this operation, and each particle has its own ideal solution (known as p_{best}) and is known as the ideal of all particles. The new numerical positions of the particle are updated using these values in each generation according to the following relationship (Eqs. 6.11 and 6.12) [28].

$$v_{ij}(t+1) = w.v_{ij}(t) + c_1 r_1 \left(pbest_{ij}(t) - x_{ij}(t)\right) + c_2 r_2 \left(gbest_{ij}(t) - x_{ij}(t)\right)$$

$$(6.11)$$

$$x_{ij}(t+1) = x_{ij}(t) + v_{ij}(t+1) \qquad (6.12)$$

where, j: 1, 2 … n, v_{ij} speed of the i particle in the iteration of j, x_{ij}, the current position of the i particle in the iteration of j, $pbest_{ij}$ is the ideal of the i particle in the iteration of j, $gbest_{ij}$ is the best position of the i particle in j iteration of the group, r_1 and r_2, is a different number between 0 and 1, c_1 and c_2 are numbers taken from 1.15 and 1.5, respectively. An essential property of PSO compared to other optimization techniques is the ability to share knowledge. In this method, the knowledge stored in the g_{best} is only transferred to others, and thus it can make it a one-way knowledge sharing system. Since evolution only finds the best solution, all particles incline to approach rapidly to the optimum state (Fig. 6.4) [28].

Case study-1

Bouacha and Terrab [29] mathematically improved the hard-turning of AISI 52100 steel using NSGA-II (non-dominated sorting genetic algorithm) and PSO-NN. It was observed that these methods had higher accuracy rate than RSM models in predicting cutting force, tool wear and surface roughness, NSGA II and PSO-NN approaches were applied efficiently and obtain similar estimation results. Furthermore, NSGA-II performed better in terms of convergence rate and precision than PSO-NN, but PSO-NN performed better in terms of calculation time. In both approaches, CS 220 m/min, FR 0.08 mm/rev, DOC 0.25 mm, cutting time 2 min and workpiece hardness 63 HRC are recommended for optimum values.

Case study-2

Jang et al. [30] modelled NN and PSO to predict the minimum cutting energy in MQL milling. It was stated that the total error between the data obtained from PSO and experimental data was less than 1%. It was also noted that the cutting energy model obtained in ANN gives accurate results in terms of input parameters (CS, DOC, FR and flow rate).

Fig. 6.4 Flow diagram of
ANN-based PSO
optimization method [28]

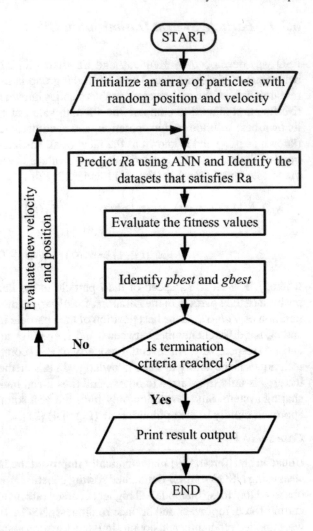

6.3.5 Gray Relational Analyses (GRA)

GRA is a MOO method that turns into multiple answers into a single-targeted
problem. It improved by Deng in 1982 to analyze uncertainties in the system and
relationships between systems [31]. In this method, the output values for interpretation and analysis are normalized between 0 and 1. These values are used to calculate
the GR coefficient of each output response. For each experimental study, the GR
grade is calculated by the mean of the GR coefficient. The higher GR grade obtained
is assumed as the ideal solution. The following stages are followed respectively for
GRA [32]. Accordingly, the output values are normalized as in Eqs. (6.13 and 6.14).
  The larger, the better for normalization;

$$x_i(k) = \frac{y_i(k) - \min y_i(k)}{\max y_i(k) - \min y_i(k)} \tag{6.13}$$

Smaller better for normalization;

$$x_i(k) = \frac{\max y_i(k) - y_i(k)}{\max y_i(k) - \min y_i(k)} \tag{6.14}$$

$x_i(k)$, is the normalized value of the output response. $\min y_i(k)$, is the smallest value of $\min y_i(k)$ for the k. answer. $\max y_i(k)$ expresses the highest value of $y_i(k)$ for k. answer. The grey relational coefficient formula is used in prediction the correlation between the ideal and the actual normalized value, in Eqs. (6.15 and 6.16).

$$\xi_i(k) = \frac{\Delta_{min} + \psi \Delta_{max}}{\Delta_{oi}(k) + \psi \Delta_{max}} \tag{6.15}$$

$$\Delta_{oi}(k) = |x_0(k) - x_i(k)| \tag{6.16}$$

ψ is the differential coefficient between 0 and 1 that extends and narrows the GR coefficient range. However, it is generally preferred as 0.5 [33]. Δ_{min}, is the least value of Δ_{oi}. Δ_{max} is the most value of Δ_{oi}. According to Eq. (6.17), the GR grade is determined in the last step of the GRA analysis (Fig. 6.5). γ_i is the GR degree and n is the number of output responses.

$$\gamma_i = \frac{1}{n} \sum_i^n \xi_i(k) \tag{6.17}$$

Case study-1

Tamrin et al. [35] optimized the laser power (LP), CS and air pressure (AP) with GRA approach for precise laser cutting of different thermoplastics. Optimum CS, LP and AP parameters for minimum HAZ (Heat affected zone) were determined as 0.4 m/min, 200 W and 2.5 bar, respectively. The low LP means low operating costs, while the high CS is similar to the optimum parameters as it allows for cutting high volumes of materials.

Case study-2

Warsi et al. [36] optimized the specific cutting energy (SCE) and material removal rate (MMR) with GRA for turning Al-6061 T6 alloy with high CSs. According to the results of the analysis, the most important parameter affecting the MO function was feed rate, and it was stated that the CS and DOC were followed, respectively. According to the proposed parameters, it was determined that SCE decreased by 5% and the MMR rate increased by 33%.

Fig. 6.5 Flow diagram of
GRA [34]

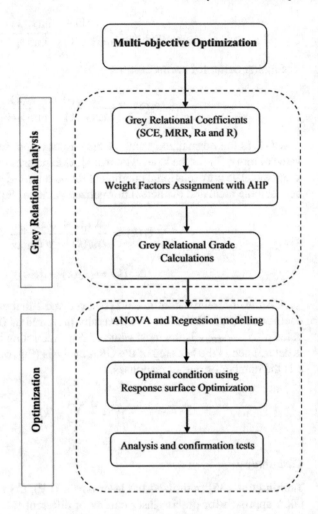

6.3.6 Technique for Order of Preference by Similarity to Ideal Solution (TOPSIS)

The TOPSIS technique was first developed by Hwang and Yoon [37]. It is an easy and multi-criteria DM method. It contributes to achieving the best optimum result among the many alternative solutions (Fig. 6.6). It is preferred to determine the best solution from the shortest from the positive ideal and the longest distance from the negative ideal. All answers are classified as useful and non-useful qualities. The valuable feature is better than the highest, whereas the less helpful feature is similar to the smaller better. The TOPSIS and DM process consist of the following stages according to the matrix system with m alternative and n criteria [38, 39].

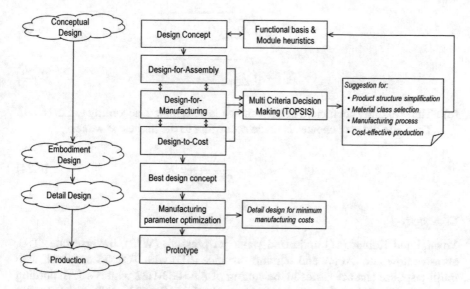

Fig. 6.6 Flow diagram for multi-objective DM TOPSIS [40]

(a) The normalized decision matrix is calculated according to Eq. (6.18). Where, y_{ij} is the normalized value and x_{kj} is the observed value in the same order as x_{ij}.

$$y_{ij} = \frac{x_{ij}}{\sqrt{\sum_{k=1}^{m} x_{ij}^2}} \tag{6.18}$$

(b) Weights are normalized within Eq. (6.19). Here, v_{ij} is the weighted normalized value, y_{ij} is the normalized value, and w_{ij} is the weight of each criterion.

$$v_{ij} = y_{ij} * w_{ij} \tag{6.19}$$

(c) Ideal (L^*) and negative ideal ($L-$) solutions are made with Eqs. (6.20 and 6.21), respectively. Here T′ is a useful criterion and T″ is the related term.

$$L^* = \{v_1^*, \ldots, v_n^*\} = \left\{ \left(\max_j v_{ij} \mathrm{li} \int T' \right), \left(\min_j v_{ij} \mathrm{li} \int T'' \right) \right\} \tag{6.20}$$

$$L^- = \{v_1^-, \ldots, v_n^-\} = \left\{ \left(\min_j v_{ij} \mathrm{li} \int T' \right), \left(\max_j v_{ij} \mathrm{li} \int T'' \right) \right\} \tag{6.21}$$

(d) The distance of each choice from the ideal positive and negative solution measures with Eqs. (6.22 and 6.23), respectively.

$$S_i^* = \sqrt{\sum_{i=1}^{n} (v_{ij} - v_i^*)^2} \qquad (6.22)$$

$$S_i^- = \sqrt{\sum_{i=1}^{n} (v_{ij} - v_i^-)^2} \qquad (6.23)$$

(e) The proximity coefficient of each choice is determined, according to Eq. (6.24). Thus, the order of choices is made according to the highest P-value.

$$P = \frac{S_i^-}{S_i^* + S_j^-} \qquad (6.24)$$

Case study-1

Yuvaraj and Kumar [41] optimized water jet pressure (WJP), traverse rate (TR), abrasive flow rate (AFR), and standoff distance (SD) with TOPSIS approach with multi-response characteristics in the cutting of AA5083-H32 with AWJ. According to the results of the analysis, optimum results for WJP, TR, AFR and SD were determined as 300 MPa, 120 mm/min, 360 g/min and 1 mm, respectively. It was stated that the most critical variable affecting the outputs was WJP, followed by TR, SD and AFR, respectively, and TOPSIS-ANOVA combination was effective for optimization.

Case study-2

Sahu et al. [42] T-SAW (Taguchi based simple additive weighting) and T-TOPSIS (Taguchi based TOPSIS) optimization approach used in turning of AA 5754 alloy. As a result of both methods, the optimum parameters were obtained as 69 m/min, 43 mm/min and 1 mm for CS, FR and DOC, respectively. It is stated that the T-SAW method is more straightforward than other Taguchi based MCDM techniques and can be used effectively for many manufacturing processes.

6.3.7 Preference Ranking Organization Method for Enrichment Evaluation (PROMETHEE)

Promethee is an MCDM technique based on binary comparisons developed by Brans [43]. It does not require the aggregation of different types of criteria values of any alternative to an absolute value. In this method, the calculation is easy and more time saving [44]. It allows ranking and selection among different criteria [45]. There are four different methods. The goal in Promethee I is full rank. While Promethee II is used to make binary comparisons between both criteria, Promethee III deals with ranges, and Promethee IV deals with numbers [46]. According to an exemplary Promethee II method, the calculation consists of four main stages [47].

Stage 1: All criteria are converted into utility criteria. Equation (6.25) and (6.26) are used for the "native" benefit and cost requirements, respectively.

$$y_{ij} = x_{ij} \tag{6.25}$$

$$y_{ij} = \max_j x_{ij} - x_{ij} \tag{6.26}$$

$i = 1, \ldots n$ and $j = 1, \ldots, j$ show the indexes of criteria and alternatives, respectively in Eq. (6.25 and 6.26).

Stage 2: Appropriate function and parameter are selected. When linear function is selected, Eq. (6.27) is taken into consideration.

$$p_i(a, b) = \begin{cases} 0, & if\ a_i - b_i \leq s_1 \\ 1, & if\ a_i - b_i > s_2 \\ \frac{a_i - b_i - s_1}{s_2 - s_1}, & otherwise \end{cases} \tag{6.27}$$

a and b represent two alternatives in Eq. (6.27). a_i and b_i represent the i. criteria values. According to the differences in classification performance, $s_1 = 0.01$ and $s_2 = 0.1$ can be considered.

Step 3: Preference index are defined and calculated for each alternative pair.

$$\begin{cases} \pi(a, b) = \sum_{i=1}^{n} p_i(a, b)w_i \\ \pi(b, a) = \sum_{i=1}^{n} p_i(b, a)w_i \end{cases} \tag{6.28}$$

Step 4: Positive and negative outranking flows are defined and calculated as in Eq. (6.29) and (6.30), respectively.

$$\phi^+(a) = \frac{1}{n-1} \sum_{x \in A, x \neq a} \pi(a, x) \tag{6.29}$$

$$\phi^-(a) = \frac{1}{n-1} \sum_{x \in A, x \neq a} \pi(x, a) \tag{6.30}$$

Stage 5: Net outranking flow for each alternative is calculated as Eq. (6.31). According to feature selection methods are listed with ϕ.

$$\phi(a) = \phi^+(a) - \phi^-(a) \tag{6.31}$$

6.3.8 Analytic Hierarchy Process (AHP)

AHP method was first improved by Saaty [48]. It divides a complex system into a system of hierarchical elements. The task of the DMr is facilitated by creating a hierarchy system with a mathematical model that generates priority values for different criteria and sub-criteria involved in the DM operation. This method uses a binary comparison of criteria at a certain level of the hierarchy to find out which rules the DMr wants to set the highest priority. These criteria are compared qualitatively. Also, some scale values corresponding to this criterion are assigned [49]. Binary comparison matrixes are created firstly to calculate the priorities for different standards. The comparison matrix is the real A matrix in the form of m × m. Where m is the number of criteria selected. Each a_{jk} entry of matrix A represents the importance of the *jth* criterion according to the *kth* standard. a_{jk} represents the entry in the *jth* line and *kth* column of matrix A [50, 51]. The relative importance between the two criteria is measured on the basis of Saaty's scale [52]. A matrix is created in the calculation of priorities and consistency check. Then, the normalized priority vector is calculated. The priority vector represents the relative weights between the criteria. The consistency of the comparison can also be measured using consistency index (CI), Random consistency index (RI) and Consistency Ratio (CR) in AHP. If CI = 0, consistency is excellent (Eq. 6.32), and if CR (Eq. 6.33) is less than 10%, it can be accepted. λ_{max} is the maximum eigenvalue and n is the size of the measured matrix in Eq. (6.32). A sample hierarchical structure for machine tool systems is given in Fig. 6.7. This structure consists of goal, category (economy, technology and ecology), criteria (tool life, machining time, cutting force, surface roughness, cutting fluid consumption, air quality and cutting power) and machining strategy alternatives (Dry, MQL and wet machining).

$$CI = \frac{(\lambda\text{max} - n)}{(n - 1)} \tag{6.32}$$

$$CR = \frac{CI}{RI} \tag{6.33}$$

6.3.9 Elimination et Choix Traduisant la REaite (ELECTRE)

ELECTRE is one of the MCDM techniques, which is based on the concept of sequencing by means of binary comparison between alternatives in the appropriate criteria. It is used for situations with many options; there are only a few criteria. It is a sequential method that reduces the number of alternative with several non-dominant options to find the best choice. It is necessary to know the weighted information of all the criteria mentioned as follow in the problem-solving step using this method [53, 54].

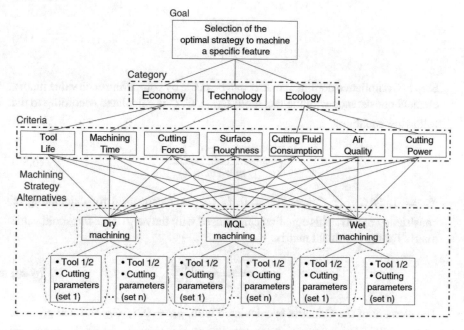

Fig. 6.7 Example hierarchical structure for machine tool systems [49]

Step 1: The decision matrix is normalized with the formula below.

$$r_{ij} = \frac{x_{ij}}{\sqrt{\sum_{i=1}^{m} x^2_{ij}}} \tag{6.34}$$

Step 2: The decision matrix normalized by the formula below is weighted.

$$V = RW \ and \ \sum_{i-1}^{n} wi = 1 \tag{6.35}$$

Step 3: Consistency and inconsistency are determined using the following function.

$$C_{kl} = \begin{cases} \{j, y_{kj} \geq y_{ij}\}, & for \ j = 1, 2, 3 \dots n \\ \{y, y_{kj} < y_{ij}\}, & for \ j = 1, 2, 3 \dots n \end{cases} \tag{6.36}$$

Step 4: The first function is used for the matrix of consistency, and the second function for the inconsistency matrix.

$$C_{kl} = \sum_{jC_w} W_j \tag{37}$$

$$d_{kl} = \frac{\max_{j \in D_{kl}} |y_{kj} - y_{lj}|}{\max_{j} |y_{kj} - y_{lj}|} \tag{38}$$

Step 5: Compliance dominant matrix can be made by comparing each value matrix element consistency with the threshold $C_{kl} \geq \underline{C}$. $\underline{C}$ is calculated according to the following equation.

$$\underline{C} = \frac{\sum_{k=1}^{n} \sum_{l=1}^{n} C_{kl}}{m * (m-1)} \tag{6.39}$$

$f_{kl} = 1$, $if\ C_{kl} \geq \underline{C}$ and $f_{kl} = 0$, $if\ C_{kl} < \underline{C}$ are used for calculating of consistency matrix. This equation can be used with the help of the threshold value matrix for the dominant matrix.

$$\underline{d} = \frac{\sum_{k=1}^{n} \sum_{l=1}^{n} d_{kl}}{m * (m-1)} \tag{6.40}$$

The element of the G matrix is calculated with, $g_{kl} = 0$, $if\ C_{kl} \geq \underline{d}$ and $g_{kl} = 1$, $if\ C_{kl} < \underline{d}$ as inconsistency of dominant matrix.

Step 6: The E matrix aggregate is found using the multiplication between the F and G matrix elements in the function below.

$$e_{kl} = f_{kl} x g_{kl} \tag{6.41}$$

Step 7: Thus, the less desirable alternative is eliminated. The aggregate domination matrix (E) gives the partial order of other options. $if\ e_{kl} = 1$ an appropriate choice is preferred for both consistency and consistency criteria.

6.4   Comparison of MCDM Techniques

Although there are many MCDM methods, a single method is not accepted as the most appropriate method for any DM method. When different MCDM methods are applied to the same problem, it is the main problem of this method that different result can be obtained. Thus, it has great importance to define and identify an appropriate MCDM method [55, 56]. It should be known that various methods can potentially be used in DM, and one technique cannot always be superior to another. One of the most important criteria in choosing MCDM technique is whether the problem is compatible with the target [55, 57]. According to some authors, it is stated that the most commonly used MCDM methods in the material selection are TOPSIS, ELECTRE and AHT [58]. It is generally stated in the literature that the AHT method is an ideal method to deal with the problems of the MCDM method [59–61]. However, when there are many alternatives, the AHT approach is not very suitable in binary comparison. In some cases, excessive calculation time restricts the use of this method. [62]. Jahan et al. [58] stated in their study that the AHT method could be evaluated in a limited number of options, not more than 15. Accordingly, it can be said that the use of AHT method will not be very effective with an increasing number of alternatives. Partial sequencing operations can be achieved with ELECTRE. As the number of alternatives increases, the volume of calculations naturally increases. The order of each alternative is determined with ELECTRE, and a numerical degree is not used to detect inconsistency between the other options [58]. While this method provides a complete order for the alternatives, it may not work efficiently due to the mathematical complexity in the selection problems. Therefore, limitations can be reduced with a more open approach by using the PROMETHEE method [63]. It is stated by Loken that the most significant difference between ELECTRE and PROMETHEE is the calculation process. While PROMETHEE has a complete calculation process, ELECTRE's calculations can be much more complicated [2]. It is a difficult process to design problems and determine weights with PROMETHEE [64]. Besides, it does not provide a perfect ranking among the options. Whereas GRA is used to achieve an ideal ranking in many problems, this aspect is similar to the TOPSIS method. The TOPSIS method uses Euclidean distance while GRA uses the gray relational grade [65]. In addition, TOPSIS is preferred more than other methods in case there are many alternatives and criteria [61]. PSO is the most straightforward optimization technique that works well in global optimization problems. TOPSIS-PSO combinations are compared with other leading methods applied in the real application-based environment. This combination increases the PSO's capacity to find optimal solutions. It also offers solutions to nonlinear two-levelled programming problems [66]. ANN technique is preferred in obtaining practical and reliable information based on the ability to learn from the environment compared to MCDM techniques, where the results are generalized in a particular area. FL technique provides many advantages in the DM process by taking into account the criteria and human decisions that cannot be measured in the evaluation process. In addition, it can provide an easy definition of the score for alternative and flexibility for expert decisions. It can be integrated with MCDM techniques such as AHP and TOPSIS to increase flexibility and sensitivity [67].

6.5 Concluding Remarks

MCDM techniques have been one of the rapidly growing research fields in recent years, depending on human needs. For the optimal solution to the problem, there is a transition from a single-criteria to a multi-criteria situation compared to the past. These criteria vary according to the working fields of the enterprises. The main objective of the optimization studies is to minimize the cost and time factor as much as possible. The more independent variables can be optimized in the criteria-based DM process, the greater the efficiency of the outputs in line with the mathematical models obtained. In order to obtain the desired output in the MCDM process, the independent variable and the appropriate objective function must be determined. For this purpose, there are many approaches to problem-solving. Among these, ANN, GRA, FL, PSO, GA and TOPSIS are widely used. It is also possible to determine the effective rates of independent variables on dependent variables with analysis of variance (ANOVA). In addition to the single-use of DM approaches, it has been determined that hybrid approaches are used in the literature. Thus, the success rate between approaches in problem-solving can be tested, and the appropriate MCDM method can be recommended for the related problem-solving. In this chapter, current case studies in the literature about optimizing the independent variables that affect the outputs such as tool wear, SR and CF in cutting theory with different MCDM approaches are discussed. It is hoped that the chapter would be a good source of knowledge for researchers and encourage further implementation of these MCDM techniques to solve a variety of commercial problems.

References

1. G.R. Jahanshahloo, F.H. Lotfi, M. Izadikhah, Extension of the TOPSIS method for decision-making problems with fuzzy data. Appl. Math. Comput. **181**(2), 1544–1551 (2006)
2. E. Løken, Use of multicriteria decision analysis methods for energy planning problems. Renew. Sustain. Energy Rev. **11**(7), 1584–1595 (2007)
3. G.A. Kiker, T.S. Bridges, A. Varghese, T.P. Seager, I. Linkov, Application of multicriteria decision analysis in environmental decision making. Integr. Environ. Asses. Manag. Int. J. **1**(2), 95–108 (2005)
4. E.D. Jato, L.E. Castillo, H.J. Rodriguez, J.J.C. Canteras, A review of application of multi-criteria decision-making methods in construction. Autom. Constr. **45**, 151–162 (2014)
5. S. Hajkowicz, A. Higgins, A comparison of multiple criteria analysis techniques for water resource management. Eur. J. Oper. Res. **184**(1), 255–265 (2008)
6. S. Dožić, Multi-criteria decision-making methods: application in the aviation industry. J. Air Trans. Manag. **79**, 101683 (2019)
7. R. Venkata Rao, B.K. Patel, Decision making in the manufacturing environment using an improved PROMETHEE method. Int. J. Prod. Res. **48**(16), 4665–4682 (2010)
8. M. Madic, J. Antucheviciene, M. Radovanovic, D. Petkovic, Determination of manufacturing process conditions by using MCDM methods: application in laser cutting. Eng. Econ. **27**(2), 144–150 (2016)
9. W. Pedrycz, P. Ekel, R. Parreiras, *Fuzzy Multicriteria Decision-Making: Models, Methods and Applications* (Wiley, 2011)
10. R.V. Rao, *Decision Making in the Manufacturing Environment: Using Graph Theory and Fuzzy Multiple Attribute Decision Making Methods* (Springer Science & Business Media, 2007)
11. S. Orak, R.A. Arapoğlu, M.A. Sofuoğlu, Development of an ANN-based decision-making method for determining optimum parameters in turning operation. Soft Comput. **22**(18), 6157–6170 (2018)
12. B. Yegnanarayana, *Artificial Neural Networks* (PHI Learning Pvt. Ltd., New Delhi, 2009)
13. F. Kara, K. Aslantas, A. Çiçek, ANN and multiple regression method-based modelling of cutting forces in orthogonal machining of AISI 316L stainless steel. Neural Comput. Appl. **26**(1), 237–250 (2015)
14. B. Das, S. Roy, R.N. Rai, S.C. Saha, Study on machinability of in situ Al–4.5% Cu–TiC metal matrix composite-surface finish, cutting force prediction using ANN. CIRP J. Manuf. Sci. Technol. **12**, 67–78 (2016)

15. D. Goldberg, *Genetic Algorithms in Search, Optimization and Machine Learning* (Addison-Wesley, Reading, MA, 1989)
16. M. Gen, R. Cheng, L. Lin, *Network Models and Optimization: Multiobjective Genetic Algorithm Approach* (Springer Science & Business Media, 2008)
17. Z. Jiang, F. Zhou, H. Zhang, Y. Wang, J.W. Sutherland, Optimization of machining parameters considering minimum cutting fluid consumption. J. Clean. Prod. **108**, 183–191 (2015)
18. A.M. Zain, H. Haron, S. Sharif, Application of GA to optimize cutting conditions for minimizing surface roughness in end milling machining process. Exp. Syst. Appl. **37**(6), 4650–4659 (2010)
19. P.K. Shrivastava, A.K. Pandey, Parametric optimization of multiple quality characteristics in laser cutting of Inconel-718 by using hybrid approach of multiple regression analysis and genetic algorithm. Inf. Phys. Technol. **91**, 220–232 (2018)
20. M. Mia, N.R. Dhar, AISI 1060 steel under effective cooling condition. Neur. Comput. Appl. **31**(7), 2349–2370 (2019)
21. V.K. Jain, An expert system for predicting the effects of speech interference due to noise pollution on humans using fuzzy approach. Exp. Syst. Appl. **35**(4), 1978–1988 (2008)
22. C.Z. Syn, M. Mokhtar, C.J. Feng, Y.H. Manurung, Approach to prediction of laser cutting quality by employing fuzzy expert system. Exp. Syst. Appl. **38**(6), 7558–7568 (2011)
23. M.V. Bobyr, S.A. Kulabukhov, Simulation of control of temperature mode in cutting area on the basis of fuzzy logic. J. Mach. Manuf. Rel **46**(3), 288–295 (2017)
24. A. Hossain, A. Hossain, Y. Nukman, M.A. Hassan, M.Z. Harizam, A.M. Sifullah, P. Parandoush, A fuzzy logic-based prediction model for kerf width in laser beam machining. Mater. Manuf. Proc. **31**(5), 679–684 (2016)
25. E. Kuram, B. Ozcelik, Micro-milling performance of AISI 304 stainless steel using Taguchi method and fuzzy logic modelling. J. Int. Manuf. **27**(4), 817–830 (2016)
26. J. Kennedy, R.C. Eberhart, Particle swarm optimization, in *IEEE International Conference on Neural Networks*, Piscataway, NJ, pp. 1942–1948 (1995)
27. I. Hanafi, F.M. Cabrera, F. Dimane, J.T. Manzanares, Application of particle swarm optimization for optimizing the process parameters in turning of PEEK CF30 composites. Proc. Technol. **22**, 195–202 (2016)
28. M. Chandrasekaran, S. Tamang, ANN–PSO integrated optimization methodology for intelligent control of MMC machining. J. Inst. Eng. (India) Ser. C **98**(4):395–401 (2017)
29. K. Bouacha, A. Terrab, Hard turning behavior improvement using NSGA-II and PSO-NN hybrid model. Int. J. Adv. Manuf. Technol. **86**(9–12), 3527–3546 (2016)
30. D.Y. Jang, J. Jung, J. Seok, Modeling and parameter optimization for cutting energy reduction in MQL milling process. Int. J. Prec. Eng. Manuf.-Green Technol. **3**(1), 5–12 (2016)
31. Y. Kuo, T. Yang, G.W. Huang, The use of grey relational analysis in solving multiple attribute decision-making problems. Comput. Ind. Eng. **55**, 80–93 (2008)
32. S. Sudhagar, M. Sakthivel, P.J. Mathew, S.A.A. Daniel, A multi criteria decision making approach for process improvement in friction stir welding of aluminium alloy. Measurement **108**, 1–8 (2017)
33. Y. Kuo, T. Yang, G.W. Huang, The use of a grey-based Taguchi method for optimizing multi-response simulation problems. Eng. Optim. **40**, 517–528 (2008)
34. M. Younas, S.H.I. Jaffery, M. Khan, M.A. Khan, R. Ahmad, A. Mubashar, L. Ali, Multi-objective optimization for sustainable turning Ti6Al4V alloy using grey relational analysis (GRA) based on analytic hierarchy process (AHP). Int. J. Adv. Manuf. Technol. **105**(1–4), 1175–1188 (2019)
35. K.F. Tamrin, Y. Nukman, I.A. Choudhury, S. Shirley, Multiple-objective optimization in precision laser cutting of different thermoplastics. Opt. Lasers Eng. **67**, 57–65 (2015)
36. S.S. Warsi, M.H. Agha, R. Ahmad, S.H.I. Jaffery, M. Khan, Sustainable turning using multi-objective optimization: a study of Al 6061 T6 at high cutting speeds. Int. J. Adv. Manuf. Technol. **100**(1–4), 843–855 (2019)
37. C.L. Hwang, K. Yoon, *Multiple-Criteria Decision Making: Methods and Applications, A State of ART Survey* (Springer, New York, 1981)

38. R. Eungkee Lee, R. Hasanzadeh, T. Azdast, A multi-criteria decision analysis on injection moulding of polymeric microcellular nanocomposite foams containing multi-walled carbon nanotubes. Plast. Rubber Compos. **46**(4), 155–162 (2017)
39. S. Gürgen, F.H. Çakır, M.A. Sofuoğlu, S. Orak, M.C. Kuşhan, H. Li, Multi-criteria decision-making analysis of different non-traditional machining operations of Ti6Al4V. Soft Comput. **23**(13), 5259–5272 (2019)
40. C. Favi, M. Germani, M. Mandolini, Development of complex products and production strategies using a multi-objective conceptual design approach. Int. J. Adv. Manuf. Technol. **95**(1–4), 1281–1291 (2018)
41. N. Yuvaraj, M. Pradeep Kumar, Multiresponse optimization of abrasive water jet cutting process parameters using TOPSIS approach. Mater. Manuf. Proc. **30**(7), 882–889 (2015)
42. A.K. Sahu, N.K. Sahu, A.K. Sahu, M.S. Rajput, H.K. Narang, T-SAW methodology for parametric evaluation of surface integrity aspects in AlMg3 (AA5754) alloy: comparison with T-TOPSIS methodology. Measurement **132**, 309–323 (2019)
43. J.P. Brans, P. Vincke, Note-a preference ranking organisation method: the PROMETHEE method for multiple criteria decision-making. Manag. Sci. **31**(6), 647–656 (1985)
44. S.P. Wan, W.C. Zou, L.G. Zhong, J.Y. Dong, Some new information measures for hesitant fuzzy PROMETHEE method and application to green supplier selection. Soft Comput., 1–25 (2019)
45. M. Behzadian, R.B. Kazemzadeh, A. Albadvi, M. Aghdasi, PROMETHEE: a comprehensive literature review on methodologies and applications. Eur. J. Oper. Res. **200**(1), 198–215 (2010)
46. M.A. Nikouei, M. Oroujzadeh, A.S. Mehdipour, The PROMETHEE multiple criteria decision-making analysis for selecting the best membrane prepared from sulfonated poly (ether ketone) s and poly (ether sulfone)s for proton exchange membrane fuel cell. Energy **119**, 77–85 (2017)
47. G. Kou, P. Yang, Y. Peng, F. Xiao, Y. Chen, F.E. Alsaadi, Evaluation of feature selection methods for text classification with small datasets using multiple criteria decision-making methods. Appl. Soft Comput. **86**, 105836 (2020)
48. T.L. Saaty, How to make a decision: the analytic hierarchy process. Eur. J. Oper. Res. **48**, 9–26 (1990)
49. O. Avram, I. Stroud, P. Xirouchakis, A multi-criteria decision method for sustainability assessment of the use phase of machine tool systems. Int. J. Adv. Manuf. Technol. **53**(5–8), 811–828 (2011)
50. A. Petruni, E. Giagloglou, E. Douglas, J. Geng, M.C. Leva, M. Demichela, Applying analytic hierarchy process (AHP) to choose a human factors technique: choosing the suitable human reliability analysis technique for the automotive industry. Saf. Sci. **119**, 229–239 (2019)
51. T.L. Saaty, *AHP: The Analytic Hierarchy Process* (McGraw-Hill, New York, USA, 1980)
52. T.L. Saaty, *Decision Making for Leaders: The Analytic Hierarchy Process for Decisions in a Complex World* (RWS Publications, Pittsburgh, USA, 1990)
53. K. Govindan, M.B. Jepsen, ELECTRE: a comprehensive literature review on methodologies and applications. Eur. J. Oper. Res. **250**(1), 1–29 (2016)
54. A. Yanie, A. Hasibuan, I. Ishak, M. Marsono, S. Lubis, N. Nurmalini, M. Mesran, S.D. Nasution, R. Rahim, A.S. Ahmar, Web based application for decision support system with ELECTRE method. J. Phys. Conf. Ser. **1028**(1), 012054 (2018)
55. E. Mulliner, N. Malys, V. Maliene, Comparative analysis of MCDM methods for the assessment of sustainable housing affordability. Omega **59**, 146–156 (2016)
56. S.H. Zanakis, A. Solomon, N. Wishart, S. Dublish, Multi-attribute decision making: a simulation comparison of select methods. Eur. J. Oper. Res. **107**, 507–529 (1998)
57. B. Roy, The outranking approach and the foundations of the ELECTRE methods. Theory Decis. **31**(1), 49–73 (1991)
58. A. Jahan, M.Y. Ismail, S.M. Sapuan, F. Mustapha, Material screening and choosing methods-a review. Mater. Des. **31**(2), 696–705 (2010)
59. M.R. Mansor, S.M. Sapuan, E.S. Zainudin, A.A. Nuraini, A. Hambali, Hybrid natural and glass fibers reinforced polymer composites material selection using Analytical Hierarchy Process for automotive brake lever design. Mater. Des. **51**, 484–492 (2013)

60. A.S. Milani, A. Shanian, C. Lynam, T. Scarinci, An application of the analytic network process in multiple criteria material selection. Mater. Des. **44**, 622–632 (2013)
61. N.S.H. Mousavi, A.A. Sotoudeh, A comprehensive MCDM-based approach using TOPSIS, COPRAS and DEA as an auxiliary tool for material selection problems. Mater. Des. **121**, 237–253 (2017)
62. P. Chatterjee, V.M. Athawale, S. Chakraborty, Materials selection using complex proportional assessment and evaluation of mixed data methods. Mater. Des. **32**(2), 851–860 (2011)
63. A. Jahan, F. Mustapha, M.Y. Ismail, S.M. Sapuan, A comprehensive VIKOR method for material selection. Mater. Des. **32**, 1215–1221 (2011)
64. M.A. Hatami, M. Tavana, M. Moradi, F. Kangi, A fuzzy group electre method for safety and health assessment in hazardous waste recycling facilities. Saf. Sci. **51**, 414–426 (2013)
65. T. Singh, A. Patnaik, R. Chauhan, Optimization of tribological properties of cement kiln dust-filled brake pad using grey relation analysis. Mater. Des. **89**, 1335–1342 (2016)
66. N. Panwar, S. Negi, M.M.S. Rauthan, K.S. Vaisla, TOPSIS–PSO inspired non-preemptive tasks scheduling algorithm in cloud environment. Clust. Comput. **22**(4), 1379–1396 (2019)
67. I. Kaya, C. Kahraman, A comparison of fuzzy multicriteria decision making methods for intelligent building assessment. J. Civil Eng. Manag. **20**(1), 59–69 (2014)

Chapter 7
Application of Multi-Criteria Decision-Making Techniques in the Optimization of Mechano-Tribological Properties of Copper-SiC-Graphite Hybrid Metal Matrix Composites

Anbesh Jamwal, Rajeev Agrawal, and Pallav Gupta

Abstract Cu-MMC with hybrid reinforcements are the attractive research area for heat exchanger, automobile and thermal management applications. The advanced properties of Cu based MMCs make them a promising choice in engineering and other material selection for many applications. There are many fabrication methods by which copper metal matrix composites can be produced, among them stir casting and powder metallurgy are famous one. In the last few years, optimization of reinforcement content, physical properties and mechanical properties of advanced composite materials have enhanced the research area of metal matrix composites. Among the optimization techniques, TOPSIS methodology is popular. In the present book chapter, Copper-SiC-Graphite hybrid metal matrix composites were prepared by liquid state stir casting process. Microstructural, Physical, Mechanical and Tribological behaviour of copper composites is investigated. It is found that the density of composites starts decreasing with increase in reinforcement content. Ultimate Tensile strength and wear resistance of fabricated copper-based composites is improved at the higher reinforcement content. TOPSIS methodology is adopted to optimize both the physical and mechanical properties of Cu-MMC. It is expected that this study will be beneficial for future research directions in the optimization of properties for metal matrix composites as well as hybrid metal matrix composites.

Keywords Metal Matrix Composites · Scanning Electron Microscopy · Mechanical Properties · TOPSIS

A. Jamwal (✉) · R. Agrawal
Department of Mechanical Engineering, Malaviya National Institute of Technology, J.L.N. Marg, Jaipur 302017, India
e-mail: 2019rme9095@mnit.ac.in

P. Gupta
Department of Mechanical Engineering, A.S.E.T., Amity University Uttar Pradesh, Noida 201313, India

© Springer Nature Switzerland AG 2021
S. Pathak (ed.), *Intelligent Manufacturing*, Materials Forming, Machining and Tribology, https://doi.org/10.1007/978-3-030-50312-3_7

7.1 Introduction

Through the ages, man has become more innovative in discovering new forms of materials to make his life more pleased and comfortable. As we know, materials are deeply seated in human's life. All the human needs such as transportation, communication, construction and even the basic daily needs depends on the materials. In the last few decades, there is a rapid change in human needs and desires which gives some advanced forms of materials with better properties from metals, non-metals, and even alloys. Materials play a very important role in human life [1]. Copper metal matrix composites are capable material to meet the present engineering demands. Copper is widely used metal all over the world, which is used in various engineering applications. The limited properties of pure metal limit its application in many engineering applications. The development of advanced MMCs can solve these problems reinforced with ceramics particulates. MMCs are widely used at present over the pure metal because of their excellent properties like higher hardness values, higher tensile strength and higher wear resistance as compared to pure metal [2]. Metal matrix composites can be fabricated by various techniques in which the stir casting fabrication process, powder-metallurgy (P/M), squeeze-casting (S/C), infiltration and In situ fabrication process are widely used to fabricate the metal matrix composites [3]. However, it is found that among all fabrication processes, stir casting fabrication process is the most economical technique for composites manufacturing. In the metal matrix composites, matrix material is of metal and reinforcement phase may be of ceramic particles [4]. The choice of reinforcement material depends on the application of composite material. In the last few years development of composites materials with the hybrid reinforcement is becoming popular because it is found that hybrid reinforcement can provide better properties such as higher wear-resistance and lower densities as compared to single reinforcement added composites [5]. Copper-based MMCs are highly promising materials for many engineering applications. Excellent thermal, electrical and ductile properties of copper metal matrix composites make them suitable choice in material selection in many engineering applications such as automobile sector, marine engineering and construction applications [6]. It is found that reinforcement content influence the properties of copper metal matrix composites. Selection of reinforcement is essential in achieving the desirable features from the composites. It is found that too much reinforcement content in the metal matrix composites can degrade the mechano-tribological properties of composites. In the past few years, developments in MMCs have made several new research areas and scopes in the field of composites materials [7]. However, these studies are limited when it comes to optimization. Optimization of reinforcement is still a research area which needs to be explored. In copper metal matrix composites generally used reinforcements are Al_2O_3, SiC, TiC, B_4C, Graphite and CNTs [8]. Development in non-traditional machining like EDM [9], Powder mixed EDM [10], Micro-EDM [11, 12] and ECM has also made new research scopes in the area of composites. In the past few years advancement in Copper metal matrix composite has taken more interest in the development of composite materials for many engineering applications [13]. In the present work Cu-MMCs reinforced with graphite and SiC particles is fabricated by Stir casting processing route. Microstructural analysis, Vickers hardness, density, tensile strength, and wear resistance of composites is investigated. Further TOPSIS technique is used to optimize the physical and mechanical properties of Cu metal matrix composites.

7.2 History of Composite Materials

Composite materials are as old as human civilization. The firstly human-made composite was introduced in 1500 B.C when the settlers of Egypt and Mesopotamia used the mixture of straw and mud for the construction of a strong building. This was the first when composite material was introduced to human society. At that time straw was used as reinforcement in the composite materials. Latterly, in the 1200 AD when Mongols were rulers of Central Asia, invention of the first composite bow took place. This bow was invented using the combination of bone, wood and animal glue. These bows were accurate and powerful as they were pressed and wrapped with birch bark. These composite bows helped Genghis Khan to ensure his military dominance. Wattle and daub, which is a composite material was used for making walls, is also the oldest man-made composite material, which is over 6000 years old. Concrete which is highly used composite material was introduced in 25 BC. With the wide range of applications in the engineering, construction and our daily life about 7.5 billion cubic meters of concrete is being made every year. The modern era of the composite materials began when humans developed the plastics. Before the development of plastics, natural resins which is derived from animals or plants were the only source of binders and glues. In the early 1900s, polyester, phenolic and plastic such as polyvinyl were developed, which made human life more comfortable and enjoyable. However, later on, human realized that plastic alone could not be able to provide the required strength for certain applications. Reinforcement was required to provide the strength needed for such kind of applications. In 1935, the first glass fibre was introduced by Owens Corning, which was a great discovery and changed the era of plastics. Because when glass fibre is combined with a plastic polymer, it creates a robust structure which is light in weight and is useful for many applications where the low weight to high strength is a requirement. This was the beginning of the FRP (Fibre-reinforced Polymer) industry. As we know, Necessity is the mother of invention. Many of developments and advancements in the composite materials were the results of wartime needs. Just as we know Mongols who have developed the composite bows for there need, same as the world war-II changed the era of Plastics. World war 2 brought the Fibre-reinforced polymer industries from the laboratory to the actual production. During the world war-II, light-weighted materials were required for military aircraft and other applications. This was the time when engineers realized some other benefits of composite materials beyond being light-weighted and strong materials for the aircraft and military applications. As the fibreglass composites were transparent and engineers realized that they could be used as Fiberglass in radio frequencies applications. Later on, these were adapted for use in electronic radar equipment's. After the World 2, the composite companies were at their peak. With the lower demands for the military products, soon the investors and company holders realized that composite materials were only limited for military equipment's. To sustained their position in the global market soon, they started introducing the lower-cost composite materials into the other markets. At that time, boats were not made up of composite, so they added the composite materials in the Boats. The first composite commercial boat was introduced in 1946. This was the time when Brandt Goldsworthy, who was a mechanical engineer from the University of California referred to as the "Father of Composites", served to many composite companies during and after the world war 2. Goldsworthy invented many manufacturing processes and products which changed the era of composites. Pultrusion process was also invented by the Goldsworthy, which allowed dependably durable fibreglass reinforced composites. In present time pipes, ladder rails, arrow shaft, armour, train floors and medical devices are manufactured from this process. Later in the 1970s again, there was a change in composite era when the industries of composite began to mature. This gave some new developments to the industry, such as better resins, improved reinforced fibres. DuPont who invented the aramid-fibre, which is also known as Kevlar, is very popular and famous body armour in the present time because of its high tensile density, high tensile strength and low weight. During this time, Carbon Fibers were also developed which replaced the parts formerly made up of steel. There is still a lot of advancements and developments which are undergoing in the composite industries to make human's life more comfortable and pleased. Plenty of developments have been done in the area of composite materials, and still, there are many gaps which need further studies.

7.3 Need of Optimization Techniques in Composite Materials

Composite materials are the most attractive candidates for present engineering applications. There is a lot of development in the fabrication and optimization techniques of composite materials in the last few years. However, there is a lack of design standards in composite materials for engineering applications which limits the use of composite materials in many applications including engineering and aerospace. Majority of the applications in composites are fabrication based in which they are fabricated by guidelines provided by manufacturers and are based on the designer or worker experience. The purpose of optimization in composite materials is to achieve the best optimal parameters and best reinforcement content to make them fit for engineering applications in the present time. There are many optimizing methods at the present time, which are using for the optimization of the reinforcement contents in the composites. Most widely used techniques are TOPSIS, Analytical hierarchy process and Genetic algorithm. Among all TOPSIS is extensively used method for optimizing the reinforcement content. The simplicity and excellent computational efficiency of TOPSIS method over the other techniques makes it a better and promising choice in the multi criteria decision making (MCDM) technique for optimizing the reinforcement contents.

7.4 Classification of Composite Materials

In the present time, composites are shaping the future of the world, which includes the fast-growing industries, aerospace and automobile sector. Due to lightweight and durable with the extended durability, they are the best alternative to the traditionally used materials. Based on matrix material, the composites materials can be categorized into four categories which are shown in Fig. 7.1 and discussed below:

A. MMCs (Metal Matrix Composites)
B. CMCs (Ceramic matrix composite)
C. PMCs (Polymer matrix composite)
D. CCMC's (Carbon-graphite matrix composite).

A. *Metal Matrix Composites*

At present, metal matrix composites (MMCs) are generating a wide range of interest in future materials, which are the best alternative over traditional materials. Metal matrix composites are the composites which have at least two constituent's parts, in which one part should be metal. The other constituent which is also called as reinforcement should be a ceramic. In the last few years, hybrid MMC's gained more popularity because of extensively better and improved properties. When at least three materials (out of which two are ceramic reinforcements) are present in Metal

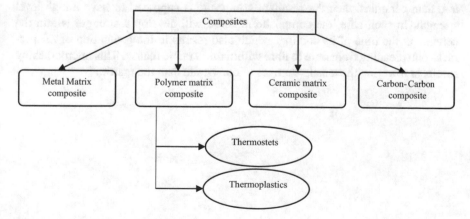

Fig. 7.1 Classification of composite based on matrix materials

matrix composite, then it is called a hybrid metal matrix composite. In the past few decades, new engineering materials have taken place in present world which fulfilled the demand of almost all the applications by maintaining the physical and mechanical properties over a longer period of time at different conditions. Rapid advancements in the miniaturization has taken place in the last few years, which makes some research scopes in the development of the lightweight MMCs. MMCs are known as new generation materials which are perfectly suitable for the present engineering needs because of their outstanding properties like higher specific strength, specific stiffness, higher wear resistance and high corrosion resistance over the traditional materials. They are having better properties over the metal, non-metals and even alloys. MMCs can be further divided into the following sub-categories:

- Al-MMCs (Aluminium metal matrix composites)
- Cu-MMCs (Copper metal matrix composites)
- Mg-MMCs (Magnesium metal matrix composites)
- Ti-MMCs (Titanium metal matrix composites)
- Fe-MMCs (Iron metal matrix composites).

B. *Ceramic Matrix Composites*

Ceramic matrix composites (CMCs) gained popularity in the past few decades due to their high melting points, good corrosion resistance strength, high compressive strength and stability at high elevated temperatures. Naturally, the ceramic matrices are an excellent choice for very high-temperature application areas. Generally, ceramics possess the low tensile strain with the high modulus of elasticity which causes the failure of adding reinforcement in the ceramic matrices to obtain strength improvement, thereby causing insufficient elongation of the matrix. This also keeps the composite from transferring the load to reinforcement. It is also found in some studies that adding the high strength fibre in the weaker ceramic matrix is not always successful; it can affect the properties of thus formed composite material. Some studies reported that when the thermal expansion of the reinforcement materials is less than ceramics, then the resultant composite is supposed to have a high-level strength. In such case, the composite material will develop a strength within the ceramic at the time of its cooling, which also results in the generation of microcracks outspreading from fibre to fibre within the ceramic matrix. This microcracking results in lowering the tensile strength of composite than the matrix.

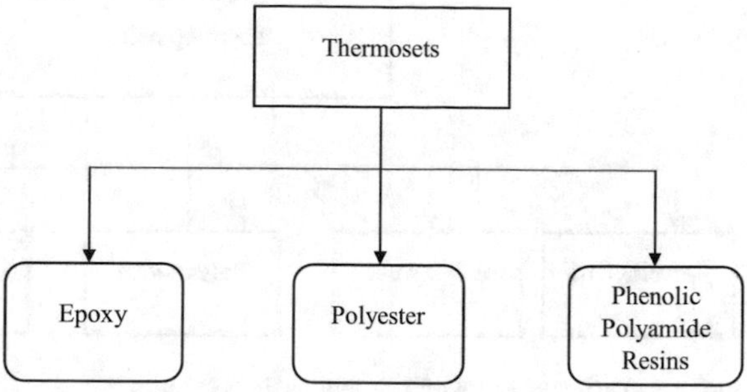

Fig. 7.2 Types
of Thermoset polymers

```
                              ┌─────────────────┐
                              │    Thermosets    │
                              └─────────────────┘


   ┌──────────────┐      ┌──────────────┐      ┌──────────────┐
   │    Epoxy     │      │  Polyester   │      │   Phenolic   │
   │              │      │              │      │  Polyamide   │
   │              │      │              │      │    Resins    │
   └──────────────┘      └──────────────┘      └──────────────┘
```

C. *Polymer Matrix composites*

Polymer matrix composites (PMCs) are the materials which can be processed easily. PMCs are lightweight materials which are the obvious choice for aeronautical applications. They possess desirable mechanical properties. Generally, there are two types of polymer materials which are Thermosets and Thermoplastics. Thermosets have the well-bonded three-dimensional structure after curing. Thermosets polymers can be categorized into three categories shown in Fig. 7.2.

D. *Carbon-carbon Composites/Carbon matrix composites*

Carbon-Carbon composites have both matrix as well as reinforcement phase of carbon. These composites having higher thermal stability, toughness and superior high speed friction properties. These composites are being produced in industries to fulfil the demands of aerospace and defence applications. In such type of composites reinforcement fibers can either be short range or continuous form having the diameter of 7-10μm. Final properties of these composites depends on the type of fabrication method used.

Thermosets are the most popular of the fibre composite matrix on which R&D of structural engineering depends. Thermosets polymers become cross-linked during the fabrication, and they don't regain their shape upon reheating. Processing of thermosets is slower than the thermoplastic polymers. They exhibit high flexibility and strength. There is better adhesion between the matrix and fibres in case of thermosets polymers. Besides this, they have better electrical properties than the thermoplastic polymers. Low shrinkage during the curing is the main advantage of thermoset polymers over the thermoplastic polymer. This helps in the production of aerospace industry parts, automobile parts, defence components etc. Epoxy matrix materials have a wide range of applications in the printed circuit board production. Thermosets have excellent electrical properties and good mechanical strength. Their processing cost is low. There are some disadvantages of thermosets like slow curing, toxicity and temperature application ranges upto a certain limit.

Thermoplastics have one or two-dimensional molecular structure and also, they can withstand up to a very high temperature. These polymers can regain their shape with heat and pressure. Their processing cost is low, and they have higher toughness as well as high volume. They have low toxicity as compared to thermoset polymers.

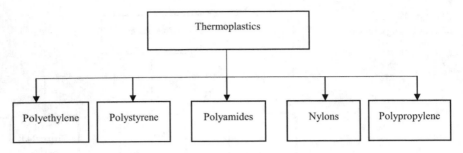

Fig. 7.3 Different types of Thermoplastic polymers

Processing of thermoplastic polymers is faster than thermoset polymers. They have low moisture absorption. Thermoplastic polymers can be categorized into five major categories shown in Fig. 7.3. In Table 7.1, different processing routes for MMCs with their applications are discussed.

7.4.1 Comparison Between Fabrication Techniques for Metal Matrix Composites

See Table 7.1.

Table 7.1 Different processing/fabrication routes for MMCs along with their application and cost

Processing route	Application	Cost
Diffusion bonding	Sheets, blades and structural components are made by diffusion Bonding	Higher investment and maintenance cost
Powder metallurgy	Mainly used in the production of small components and mass production	Investment and maintenance cost is medium
Squeeze casting	Wide range of applications in the automotive industry mainly used in the production of piston	Investment and maintenance cost is medium
Liquid metal infiltration	Used to produce structural engineering shapes and rods	The investment cost is low, and the maintenance cost is medium
Spray casting	Used in the production of electrical brushes and grinding tools	Investment and maintenance cost is medium
Compo casting or rheo-casting	Wide range of use in the production of industrial equipment and sporting goods	Both the investment and maintenance cost is low

7.4.2   Challenges with Composites Materials at Present

1. Raw materials available at a very higher cost and fabrication is also costly.
2. Generally, Composites are brittle so they can damage easily.
3. Traverse properties of composite materials are weak.
4. Since the matrix is weak in the composite materials, and due to this, they may have low toughness.
5. Analysis of composite is very difficult.
6. Reuse and disposal of the composites are difficult.
7. Resins used in composite materials weaken at very low temperatures like 150 °C so it is difficult to use those types of composite in the aerospace industry because it can produce toxic fumes on burning which may be harmful to the humans.
8. They are more expensive than traditional materials.
9. They are hard in nature, so if there will be interior damage, then it would be difficult to locate.
10. Repairs are difficult for composite materials.

7.5   Literature Review

In the last few years, developments in metal matrix composites have made several research scopes which have led to new research directions in the development of advanced metal matrix composites. Zhou et al. (2019) investigated the effect of ZrO_2 crystal structure on the wear properties of Cu-MMC, and it is found that microparticles of ZrO_2 have lower hardness and surface roughness in micro friction tests. It is also found that ZrO_2 particles improve the wear resistance of copper composites [14]. Zhou et al. (2019) investigated both wear maps and friction coefficient of Cu-MMCs with the different iron volume content. In the study, 3D friction and wear maps is constructed and assessed to analyse the wear rate and wear mechanism of composites. It is found that 10% and 15% volume content exhibited a higher coefficient of friction [15]. Salvo et al. (2019) investigated both electrical and mechanical properties of graphene reinforced copper metal matrix composites and concluded that there was a 22% rise in electrical conductivity of copper after the dispersion of graphite in the matrix phase. Better interfacial bonding of graphite and copper reduces the wear rate of composites [16]. Prajapati and Chaira (2019) fabricated the Cu-B_4C metal matrix composites by solid-state sintering and investigated its effect on microstructure, electrical and mechanical properties. It is found that with an increase in boron carbide reinforcement contents, Vickers hardness of composites increases. The maximum relative density of 82% is achieved for the 5 wt% of reinforcement. There is a decrease in electrical conductivity of composites on the dispersion of boron carbide as reinforcement [17]. Gong et al. (2019) investigated the structural and mechanical behaviour of Cu-SiO_2 and Cu-Cr metal matrix composites. It is found that both the reinforcement phases were broken during the sliding test due to the weak interfacial bonding between the matrix and reinforcement phases. XRD results reveal that there is no interface reaction occuring between the matrix and reinforcement phase [18]. Liu et al. (2019) investigated the effect of thermal cycling on the CNT reinforced Cu-MMCs by powder metallurgy process. Mechanical properties of the materials were also studied in the temperature range of 196–200 °C. It is found that the tensile strength of the composites decreases due to the increase in thermal cycles. Generally, an increase in thermal cycles results in the formation of micro-voids, which further led to the establishment in cracks. Also, the dislocations in the Cu matrix is found due to the increase in thermal cycles [19]. Jabinth and Selvakumar (2019) discussed the effect of vanadium reinforcements on the mechanical behaviour of Copper metal matrix composites fabricated by stir casting route. It is found that with the addition of vanadium reinforcement there is an increase in the strength of matrix. Improvement in hardness is seen in the elements with an increase in reinforcement content. Wear rate of composites is also decreasing due to the strong bonding between the matrix and reinforcements [20]. Su et al. (2019) studied the effect of grinding mechanism of brake pad with the Cu metal matrix composites for railways applications. It is found that processing of Cu metal matrix composites is still a challenging task for researchers in which uniform distribution of reinforcement particles over the matrix is yet a significant area for the researchers. In the present study effect of graphite particles is investigated in the Cu matrix and found that brittle fracture of graphite particles led to the plastic formation, which makes the Cu not for form the chips. The brittle fracture is the main cause of removal of the brake pad materials in railway applications [21]. Khobragade et al. (2019) investigated the electrical and mechanical behaviour of Cu-Graphene nanocomposites processed by high-pressure torsion technique. It is found that Cu-10 wt% graphene composite shows a higher hardness value and electrical conductivity. TEM investigation reveals that there is a stronger interface between the graphene and Cu which acts as a barrier for dislocations [22]. In the next section of chapter processing of Cu-SiC-graphite hybrid metal matrix composites are discussed.

7.6 Challenges and Research Gaps in Copper-MMCs

From the above-presented literature, it is found that there is a need to optimize the reinforcement content in Copper composites to achieve the best properties. Too much reinforcement contents in the matrix material can degrade the mechanical properties of the materials. It also affects the physical and structural properties of the material. Based on the literature presented, the following research questions can be drawn:

RQ1.   What is the need for optimization in copper composites?
RQ2.   What is the optimal reinforcement content for Copper composites?
RQ3.   Which method can be used to optimize the reinforcement content for Copper composites?

To answer the above research question following objectives are set:

1. Optimization in composites helps to achieve the best properties, which include the structural, physical and mechanical. The fabrication can be done with the help of stir casting route, which is economical and widely used in many industries.
2. The physical and mechanical properties are investigated based on different reinforcement contents. Each test is repeated for 5 times to avoid any type of error in the study.
3. TOPSIS method is used to optimize the reinforcement content because of its simplicity and excellent computational efficiency.

7.7 Experimentation

In this research work, Cu based MMCs reinforced with SiC and graphite are fabricated by stir casting. Total Five compositions were prepared which are shown in Table 7.2. For each composition 02 set of samples were fabricated. Casting furnace with a PID controller is used as a melting furnace. Graphite crucible is used to melt and extract the molten metal. Mechanical stirrer with SiC base material is used as stirring material. Deoxidized High-phosphorus copper (C10800 grade) is used as the matrix material. The composition of Cu-DHP is shown in Table 7.3. Graphite powder with 300 mesh size (Purity = 99%) and SiC powder with 300 mesh size (Purity = 99%) are used as the reinforcement phase materials to act as a dispersant in the composites. Copper-DHP ingots were put into the graphite crucible, and temparature of PID controlled furnace was raised to 1100 °C. The copper ingots were converted into molten form after 1100 °C. In order to ensure proper uniformity in melt, furnace temperature was raised to 1150 °C. Reinforcement materials were preheated in a separate muffle furnace at 800 °C for 60 min in order to avoid oxidation. Then, reinforcements were added in the molten metal and stirring action was done with a mechanical stirrer at 400 rpm for 45 min. After the uniform mixing of reinforcement content into molten metal, slurry was poured into preheated mould

Table 7.2 Composition of composite samples

S. No.	Sample code	Copper (wt%)	SiC (wt%)	Graphite (wt%)
1	SGC01	100	0	0
2	SGC02	99	0.5	0.5
3	SGC03	98	1	1
4	SGC04	97	1.5	1.5
5	SGC05	96	2	2

Table 7.3 Composition of Deoxidized High-phosphorus copper

Copper	Phosphorus (Min.)	Phosphorus (Max.)
99.95%	0.005%	0.013%

and allowed to solidify. Further, lathe machine was used to cut samples for testing purposes.

7.8 Microstructural and Physical Property Measurements

7.8.1 Scanning Electron Microscopy

SEM of prepared hybrid SiC-Graphite reinforced copper metal matrix composite is done at different magnification ranges to ensure the dispersion of SiC and Graphite particles in the copper matrix.

7.8.2 Density Measurement

It is the characteristic property of any object. In simple, Density is the relationship between the mass of a particular object and space (Volume) taken by its weight. The density of any object depends upon their atomic mass, size of atoms and arrangement of atoms. The Archimedes Principle determines the density of the composite. However, some error may occur during the measurement of density because of pores in the composite material. There are some steps which are used to measure the density of the composite, which are:

- Measure the dry weight of the composites.
- Composite dipped in the water for a fixed period (24 h) so that the pores between the composites can be filled up with water.
- Dipped weight of composite is measured
- Soaked weight of composite is measured.

After the measurements, densities of the composites can be determined by the given formula in Eq. (7.1).

$$\begin{aligned}
\text{Experimental Density of composite} \\
= \frac{\text{Dry weight of composite}}{\text{Soaked weight of the composite} - \text{Dipped weight of the composite}} \\
\times \text{Density of water}
\end{aligned} \qquad (7.1)$$

7.9 Mechanical Property Measurements

7.9.1 Hardness

The hardness of SiC-Graphite Reinforced Copper metal matrix composite is measured by Vickers hardness tester. To get an optimum result minimum of five readings was taken from each sample and then average from the five readings is considered as a final result.

7.9.2 Ultimate Tensile Strength

UTS test for the copper-based MMCs reinforced with SiC-Graphite is conducted on a universal testing machine. ASTM standards were followed for the sample preparation and testing purposes.

Further, TOPSIS technique is used to optimize the reinforcement content in the composites, which can give the optimal properties.

7.9.3 Wear Test

Wear test is conducted on Pin-on-Disc wear tester. Specimen dimensions for wear test were 10×25 mm. Wear test was conducted on four different loads for a fixed time of 60 min. The formula for the calculation of RPM was:

$$\frac{\pi d n}{60 \times 1000} = 4 \qquad (7.2)$$

The load applied for wear test was 10 N, 20 N, 30 N and 40 N for a fixed time of 60 min.

7.10 Results and Discussion

7.10.1 Microstructural Analysis

SEM micrographs for the composites SGC02, SGC03, SGC04 and SGC05 are shown in Fig. 7.4. It is found that uniform dispersions of reinforcement particles in matrix material takes place due to the better wettability between matrix and reinforcement particles. SEM micrographs were taken at 10.00 KX magnification range. The uniform dispersion of reinforcement particles is responsible for better physical and

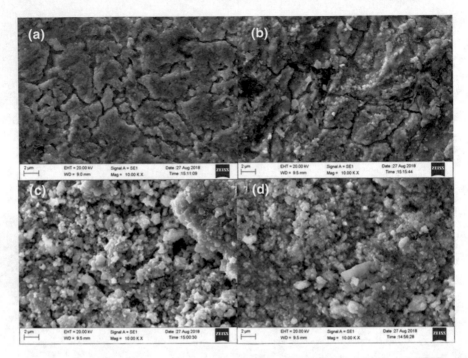

Fig. 7.4 Scanning electron microscopic images of samples (**a**) SGC02, (**b**) SGC03, (**c**) SGC04 and (**d**) SGC05 at 10.00KX

mechanical properties of the material. It is also found that there is some agglomeration effect which is seen at higher reinforcement content because the mixing of graphite in the Cu metal matrix is a difficult task by stir-casting technique [23]. At the lower reinforcement content mixing is uniform but at higher reinforcement content, there was difficulty in mixing of reinforcement content due to higher density variation and lower wettability between the matrix and reinforcement material.

7.10.2 Density and Hardness

The density of composites is measured by calculating its mass and volume, which is shown in Fig. 7.5. It is found that the density of composites starts decreasing on the addition of reinforcement particles. Generally, ceramic particles possess lower densities, and when they are added as reinforcement, they lower the density of the material. Similarly, incorporation of graphite particles as reinforcement material lowers the density of composite because of the lower density of graphite [24]. It is found that maximum density is 8.96 g/cc for sample SGC01 and on the addition of reinforcement particles density reduces to 8.9 g/cc for sample SGC02. Minimum

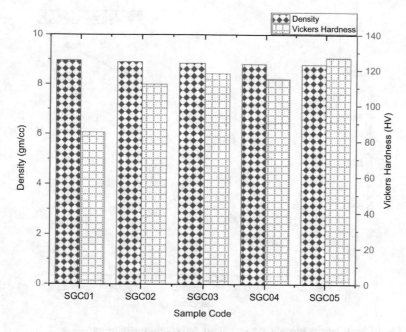

Fig. 7.5 Density (gm/cc) and Vickers Hardness (HV) graph for composites

density is found for sample SGC05 because of higher reinforcement content of SiC and Graphite which lowers the density of composites.

The hardness of prepared composites is investigated on Vickers-Hardness tester. The hardness of Cu as well as Cu-MMCs samples is shown in Fig. 7.5. It is found that when there is an increase in the reinforcements content hardness of composites starts increasing. There is a sudden reduction in hardness of composites as graphite content is increased in reinforcement because it is found that composites reinforced with graphite deform easily during the mechanical characterization. The crack propagation easily grows in case of graphite-reinforced composites [25]. However, the equal amount of SiC in reinforcement balances the reduction of hardness. It is found that maximum hardness is found for the SGC05 sample which is 127 HV and for sample SGC02 when there is the incorporation of graphite particles as reinforcement then the hardness of sample starts increasing due to better interfacial bonding between the matrix and reinforcement phases.

7.10.3 Ultimate Tensile Strength

Figure 7.6. shows the ultimate tensile strength (UTS) of the composites. It is found that there is an increase in the ultimate tensile strength (UTS) of samples on increasing the reinforcements content. It is found that there is an increase in tensile strength by

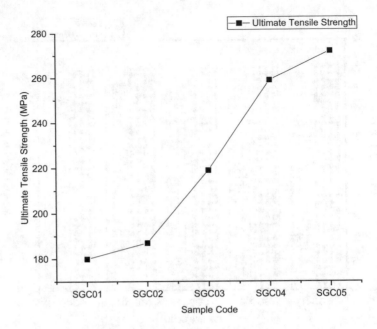

Fig. 7.6 Ultimate tensile strength of pure copper and composite samples

27% for the composite samples. The ultimate tensile strength in samples is increased due to better interfacial bonding between the Cu and reinforcement particles. Wettability is the dominant factor which promotes the better interfacial bonding between the reinforcement and matrix material [26]. Good interfacial bonding between the Cu and SiC-Graphite results in an improvement in tensile strength of samples.

7.10.4 Wear Resistance

Wear-tests were performed on samples using Pin-on-Disc wear testing machine at average room temperature without any external lubrication. Firstly, the disc was cleaned with acetone to avoid any interruption in wear test. The load tests were performed in the load range of 10 N, 20 N, 30 N and 40 N. Figure 7.7. shows the wear rate of pure copper and different composite samples at different loading conditions. It is found that wear-resistance of composites is higher at the higher reinforcements content. This is due to the proper and uniform mixing of reinforcement particles. It is seen that ceramic particles are hard, and they resist the deformation mechanism during the wear analysis. Similarly, graphite particles are soft, and they provide a lubrication layer on the counterpart [27]. Increase in lubrication during the wear analysis decreases the wear rate of samples. Hence, wear resistance of material increases. It is also found that wear rate is maximum for pure Cu-DHP sample, and

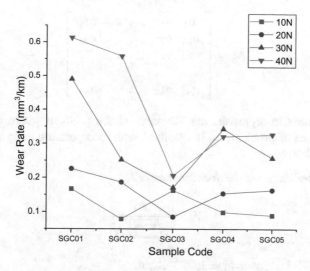

Fig. 7.7 Wear rate of pure copper and composite samples at different loads

when the reinforcement content starts increasing, there is gradually less loss in wear. Maximum wear resistance is found for the sample SGC05.

7.11   TOPSIS Optimization

Optimization of physical and mechanical properties of composites is an emerging research area at the present time [28, 29]. Optimization of properties can be done through TOPSIS [30], analytical hierarchy process [31, 32] and hybrid techniques such as Fuzzy-TOPSIS [33] and Fuzzy-Analytical hierarchy process [34]. TOPSIS method is one of the popular ways among all multi-criteria decision-making techniques for selecting the best alternatives in available options [35, 36]. The fundamental idea of TOPSIS methodology does not only provide the lowest distance from the best optimal solution of the problem but also provides the largest distance from the worst solution of the problem [37]. These days optimization with the TOPSIS is becoming popular as the development in composites materials have made new research scopes with the new reinforcements, so there is need to optimize the properties of composites materials with reinforcement content ratio [38, 39]. The steps to be followed in TOPSIS optimization techniques is discussed below:

Step 1: Formulation of Decision Matrix of the problem:

$$A_{ij} = \begin{bmatrix} a_{11} & a_{12} & \cdots & \cdots & a_{1n} \\ a_{21} & a_{22} & \cdots & \cdots & a_{2n} \\ \cdots & \cdots & \cdots & \cdots & \cdots \\ \cdots & \cdots & \cdots & \cdots & \cdots \\ a_{m1} & a_{m2} & \cdots & \cdots & a_{mn} \end{bmatrix} \tag{7.3}$$

In the matrix, n represents the attributes of the problem where m represents the alternatives of problem. A_{ij} is specified with incorporating every attribute and alternative of problem.

Step 2: Normalization of the decision matrix

$$F_{ij} = \frac{A_{ij}}{\sqrt{\sum_{i=1}^{m} A_{ij}^2}} \tag{7.4}$$

where i = 1, 2, 3, ...m and j = 1, 2, 3, ... n.

Step 3: Weighted normalized decision matrix calculation

$$b_{ij} = c_i V_{ij} \quad i = 1, 2, 3 \ldots m \text{ and } j = 1, 2, 3 \ldots n \tag{7.5}$$

Step 4: Positive and Negative ideal solution calculation

Based on normalized weighted rank (b_{ij}) the ideal X^+ and ideal X^-, solution can find out as follows:

$$X^+ = \left(b_1^+, b_2^+, b_3^+ \ldots b_4^+ \right) \tag{7.6}$$

$$X^- = \left(b_1^-, b_2^-, b_3^- \ldots b_4^- \right) \tag{7.7}$$

$$b_j^+ \begin{cases} \max_i b_{ij} \text{ if j, benefit attribute} \\ \\ \min_i b_{ij} \text{ if j, cost attribute} \end{cases}$$

$$b_j^- \begin{cases} \min_i b_{ij} \text{ if j, benefit attribute} \\ \\ \max_i b_{ij} \text{ if j, cost attribute} \end{cases}$$

Step 5: Calculation of distance with an ideal solution

With a positive ideal solution distance is an alternative A_i which is supposed as follows:

$$D_i^+ = \sqrt{\sum_{j=1}^{n} (b_{ij} - b_j^+)} \quad i = 1, 2, 3, \ldots m \qquad (7.8)$$

With a negative ideal solution distance is an alternative X_i which is supposed as follows:

$$D_i^- = \sqrt{\sum_{j=1}^{n} (b_{ij} - b_j^-)} \quad i = 1, 2, 3, \ldots m \qquad (7.9)$$

Step 6: Calculation of the preference value

For every alternative (V_i), the preference value has been given as:

$$V_i = \frac{D_i^-}{D_i^- + D_i^+} \quad i = 1, 2, 3, \ldots m \qquad (7.10)$$

The greater value of V_i indicates that alternative X_i is preferred, at the end of the calculation.

7.12 Optimization of Physical and Mechanical Properties of Copper-SiC-Graphite Hybrid Metal Matrix Composites by Using TOPSIS Method

Based on non-beneficiary and beneficiary attributes formulation of the decision matrix is done in Table 7.4.

Table 7.4 represents the attributes selected for the present study with their respective selection criteria. There are two types of attribute chosen for the study, i.e. non-beneficiary attributes and beneficiary attributes. The density of composite is

Table 7.4 Attributes with their respective selection criteria	S. No.	Attributes	The selection criterion of attributes
	1	Density	Non-beneficiary attribute (lower the better)
	2	Vickers Hardness (VH)	Beneficiary attribute (higher the better)
	3	Ultimate tensile strength (UTS)	Beneficiary attribute (higher the better)

considered as a non-beneficiary attribute, which should be lower. The present needs of engineering applications are lower densities, so the density is considered as a non-beneficiary attribute. Hardness and UTS of composites should be higher for the engineering applications, so both of these are considered as beneficiary attributes.

According to Eq. (7.3), as depicted in TOPSIS methodology, decision matrix for the given problem is formed. In the matrix, a row represents the alternatives, and the column represents the attributes of the problem. Five reinforcement contents as alternatives with their four attributes are shown in the matrix. In the matrix except for density, all the attributes are beneficiary.

Table 7.5 discusses the decision matrix considered for the present problem, the data for the decision matrix is taken from the results obtained from the characterization of composites. The values have been shown with their respective sample codes.

Table 7.6 represents the square root for the decision matrix of the problem, which is further required for normalizing the matrix as shown in Table 7.7. Table 7.8 shows the positive ideal solutions for problem of all reinforcement contents. In Table 7.8, D_i^+ represents the positive ideal solution of the problem.

Table 7.8 shows the negative ideal solutions for the problem of all reinforcement contents. In the Table 7.9 Di^- represents the negative ideal solution of the problem. The preference values for all the reinforcement contents is shown in the Table 7.10. Here Ci represents the preference value for all reinforcement contents. Table 7.11 represents the ranking order for the reinforcement content in the ascending order. In this ranking SGC02 has the optimal reinforcement content for Cu-MMCs.

The optimization of reinforcement contents in the Copper-SiC-Graphite hybrid metal matrix composites is done by TOPSIS technique. The preference values of

Table 7.5 Decision matrix of the problem

	Density (gm/cc)	Vickers Hardness (HV)	Ultimate Tensile strength (MPa)
SGC01	8.96	85	180
SGC02	8.9	112	187
SGC03	8.85	118	219
SGC04	8.81	115	259
SGC05	8.79	127	272

Table 7.6 Square root for the decision matrix of the problem

SGC01	80.2816	7225	32,400
SGC02	79.21	12,544	34,969
SGC03	78.3225	13,924	47,961
SGC04	77.6161	13,225	67,081
SGC05	77.2641	16,129	73,984
Sum	392.6943	63,047	256,395
Root SQ	19.81651584	251.0916167	506.3546188

Table 7.7 Normalized matrix of the problem

Normalized matrix table		
SGC01	**0.452148101**	0.338521855	0.355482094
SGC02	0.449120323	0.446052327	0.369306397
SGC03	0.446597175	0.469947988	0.432503214
SGC04	0.444578657	0.458000157	0.511499235
SGC05	0.443569398	**0.505791478**	**0.537172941**

Table 7.8 Calculation for D_i^+

				Sum	D_i^+
SGC01	0	0.027979127	0.033011564	0.060990691	0.246963
SGC02	9.16744	0.003568766	0.028179177	0.03175711	0.178205
SGC03	3.08128	0.001284756	0.010955752	0.01227132	0.110776
SGC04	5.72965	0.00228401	0.000659139	0.003000446	0.054776
SGC05	7.35941	0	0	7.35941	0.008579

Table 7.9 Calculation for D_i^-

				Sum	D_i^-
SGC01	7.35941	0	0	7.35941	0.008579
SGC02	3.08128	0.011562802	0.000191111	0.011784726	0.108557
SGC03	9.16744	0.017272828	0.005932253	0.023214249	0.152362
SGC04	1.0186	0.014275065	0.024341348	0.038617432	0.196513
SGC05	0	0.027979127	0.033011564	0.060990691	0.246963

Table 7.10 Calculation of Ci

	Ci	
SGC01	0.043315506	5
SGC02	1.178205248	1
SGC03	1.110775992	2
SGC04	1.054776327	3
SGC05	1.008578703	4

Table 7.11 Ranking order for reinforcement

SGC02	1
SGC03	2
SGC04	3
SGC05	4
SGC01	5

fabricated composite materials have been arranged by giving priority to the higher values. The ranking order obtained after optimization shows that SGC02 composite has an optimum result for the physical and mechanical properties of Copper-SiC-Graphite hybrid metal matrix composite.

7.13 Conclusion

Copper-SiC-Graphite hybrid metal matrix composites was successfully fabricated by liquid state stir casting technique. Based on the results obtained from the analysis, following conclusions can be drawn:

1. Microstructural analysis of composites shows the uniform distribution of reinforcement particles in the Cu matrix. However, at higher reinforcement content, some agglomeration effect is seen due to higher density variation between reinforcement and matrix material.
2. It is found that the density of composites starts decreasing on increasing the reinforcement content, which is due to the lower densities of reinforcement particles.
3. The hardness of composites is improved due to the uniform dispersion of SiC particles which resist the plastic deformation during the mechanical characterization of composites.
4. Ultimate tensile strength of composites is improved due to uniform distribution and better interfacial bonding of reinforcement particles within the matrix material.
5. Wear resistance of composites starts increasing with an increase in reinforcement content; this is because graphite provides a lubrication layer on the counterpart, which reduces the wear rate of composites.
6. TOPSIS analysis shows that SGC02 composite provides the optimum physical and mechanical properties among all the samples. It is found that with this composition industries can achieve the best properties for engineering applications.

Acknowledgements This research work is sponsored by TEQIP-3 sponsored project "Exploration of sustainable manufacturing for Indian Manufacturing industries".

References

1. A. Jamwal, U.K. Vates, P. Gupta, A. Aggarwal, B.P. Sharma, Fabrication and characterization of Al_2O_3–TiC-reinforced aluminium matrix composites, in *Advances in Industrial and Production Engineering* (Springer, Singapore, 2019), pp. 349–356

2. P. Garg, A. Jamwal, D. Kumar, K.K. Sadasivuni, C.M. Hussain, P. Gupta, Advance research progresses in aluminium matrix composites: manufacturing & applications. J Mater Res Technol **8**, 4924–4939 (2019)

3. S. Hossain, M.M. Rahman, A. Jamwal, P. Gupta, S. Thakur, S. Gupta, Processing and characterization of pine epoxy-based composites, in *AIP Conference Proceedings*, vol. 2148, no. 1 (AIP Publishing, 2019), p. 030017

4. A. Kumar, M.Y. Arafath, P. Gupta, D. Kumar, C.M. Hussain, A. Jamwal, Microstructural and mechano-tribological behavior of Al reinforced SiC-TiC hybrid metal matrix composite. Mater. Today Proc. **21**, 1417–1420 (2020)

5. S.T.I. Nayim, M.Z. Hasan, P.P. Seth, P. Gupta, S. Thakur, D. Kumar, A. Jamwal, Effect of CNT and TiC hybrid reinforcement on the micro-mechano-tribo behaviour of aluminium matrix composites. Mater. Today Proc. **21**, 1421–1424 (2019)

6. K. Bandil, H. Vashisth, S. Kumar, L. Verma, A. Jamwal, D. Kumar, N. Singh, K. K. Sadasivuni, P. Gupta, Microstructural, mechanical and corrosion behaviour of Al–Si alloy reinforced with SiC metal matrix composite. J. Compos. Mater. **53**(28–30), 4215–4223 (2019)

7. S. Hossain, M.M. Rahman, D. Chawla, A. Kumar, P.P. Seth, P. Gupta, D. Kumar, R. Agrawal, A. Jamwal, Fabrication, microstructural and mechanical behavior of Al-Al$_2$O$_3$-SiC hybrid metal matrix composites. Mater. Today Proc. **21**, 1458–1461 (2019)

8. M.A.Z. Sohag, P. Gupta, N. Kondal, D. Kumar, N. Singh, A. Jamwal, Effect of ceramic reinforcement on the microstructural, mechanical and tribological behavior of Al-Cu alloy metal matrix composite. Mater. Today Proc. **21**, 1407–1411 (2020)

9. S. Goyal, A. Jamwal, R. Pandey, Optimization of process parameters in electro-discharge machining using Taguchi method. Carbon **100**, 0–14 (2016)

10. S.T.I. Nayim, M.Z. Hasan, A. Jamwal, S. Thakur, S. Gupta, Recent trends & developments in optimization and modelling of electro-discharge machining using modern techniques: a review, in *AIP Conference Proceedings*, vol. 2148, no. 1 (AIP Publishing, 2019), p. 030051

11. A. Jamwal, A. Aggarwal, N. Gautam, A. Devarapalli, Electro-discharge machining: recent developments and trends. Int. Res. J. Eng. Technol. **5**, 433–448 (2018)

12. K. Kakkar, N. Rawat, A. Jamwal, A. Aggarwal, Optimization of surface roughness, material removal rate and tool wear rate in EDM using Taguchi method. Int. J. Adv. Res. Ideas Innov. Technol. **4**, 16–24 (2018)

13. A. Jamwal, P. Mittal, R. Agrawal, S. Gupta, D. Kumar, K.K. Sadasivuni, P. Gupta, Towards sustainable copper matrix composites: manufacturing routes with structural, mechanical, electrical and corrosion behaviour. J. Compos. Mater., DOI: https://doi.org/10.1177/002199831 9900655 (2020)

14. H. Zhou, P. Yao, T. Gong, Y. Xiao, Z. Zhang, L. Zhao, K. Fan, M. Deng, Effects of ZrO$_2$ crystal structure on the tribological properties of copper metal matrix composites. Tribol. Int. (2019)

15. H. Zhou, P. Yao, Y. Xiao, K. Fan, Z. Zhang, T. Gong, L. Zhao, M. Deng, C. Liu, P. Ling, Friction and wear maps of copper metal matrix composites with different iron volume content. Tribol. Int. **132**, 199–210 (2019)

16. C. Salvo, R.V. Mangalaraja, R. Udayabashkar, M. Lopez, C. Aguilar, Enhanced mechanical and electrical properties of novel graphene reinforced copper matrix composites. J. Alloy. Compd. **777**, 309–316 (2019)

17. P.K. Prajapati, D. Chaira, Fabrication and characterization of Cu–B$_4$C Metal matrix composite by powder metallurgy: effect of B$_4$C on Microstructure, mechanical properties and electrical conductivity. Trans. Indian Inst. Met. **72**(3), 673–684 (2019)

18. T. Gong, P. Yao, X. Xiong, H. Zhou, Z. Zhang, Y. Xiao, L. Zhao, M. Deng, Microstructure and tribological behavior of interfaces in Cu-SiO$_2$ and Cu-Cr metal matrix composites. J. Alloy. Compd. **786**, 975–985 (2019)

19. J. Liu, D.B. Xiong, Y. Su, Q. Guo, Z. Li, D. Zhang, Effect of thermal cycling on the mechanical properties of carbon nanotubes reinforced copper matrix nanolaminated composites. Mater. Sci. Eng., A **739**, 132–139 (2019)

20. J. Jabinth, N. Selvakumar, Effect of Vanadium on enhancing the mechanical and wear behaviour of copper by using stir casting technique. Mater. Res. Express (2019)

21. C. Su, C. Wang, X. Sun, X. Sang, Study on grinding mechanism of brake pad with copper matrix composites for high-speed train. Adv. Mater. Sci. Eng. (2019)
22. N. Khobragade, K. Sikdar, B. Kumar, S. Bera, D. Roy, Mechanical and electrical properties of copper-graphene nanocomposite fabricated by high pressure torsion. J. Alloy. Compd. **776**, 123–132 (2019)
23. A. Jamwal, P. Prakash, D. Kumar, N. Singh, K.K. Sadasivuni, K. Harshit, S. Gupta, P. Gupta, Microstructure, wear and corrosion characteristics of Cu matrix reinforced SiC–graphite hybrid composites. J. Compos. Mater. **53**(18), 2545–2553 (2019)
24. N. Radhika, R. Karthik, S. Gowtham, S. Ramkumar, Synthesis of Cu-10Sn/SiC metal matrix composites and experimental investigation of its adhesive wear behaviour. Silicon **11**(1), 345–354 (2019)
25. A. Jamwal, P.P. Seth, D. Kumar, R. Agrawal, K.K. Sadasivuni, P. Gupta, Microstructural, tribological and compression behaviour of Copper matrix reinforced with Graphite-SiC hybrid composites. Materials Chemistry and Physics, 123090 (2020)
26. H.J. Cho, D. Yan, J. Tam, U. Erb, Effects of diamond particle size on the formation of copper matrix and the thermal transport properties in electrodeposited copper-diamond composite materials. J. Alloy. Compd. **791**, 1128–1137 (2019)
27. L. Meng, X. Wang, X. Hu, H. Shi, K. Wu, Role of structural parameters on strength-ductility combination of laminated carbon nanotubes/copper composites. Compos. A Appl. Sci. Manuf. **116**, 138–146 (2019)
28. R. Matsuzaki, R. Yokoyama, T. Kobara, T. Tachikawa, Multi-objective curing optimization of carbon fiber composite materials using data assimilation and localized heating. Compos. A Appl. Sci. Manuf. **119**, 61–72 (2019)
29. R.Q. Sardinas, P. Reis, J.P. Davim, Multi-objective optimization of cutting parameters for drilling laminate composite materials by using genetic algorithms. Compos. Sci. Technol. **66**(15), 3083–3088 (2006)
30. S. Ramesh, R. Viswanathan, S. Ambika, Measurement and optimization of surface roughness and tool wear via grey relational analysis, TOPSIS and RSA techniques. Measurement **78**, 63–72 (2016)
31. N. Gautam,, M.K. Ojha, P. Swain, A. Aggarwal, A. Jamwal, Informal investigation of fourth-party and third-party logistics service providers in terms of Indian context: an AHP approach, in *Advances in Industrial and Production Engineering*(Springer, Singapore, 2019), pp. 405–413
32. P. Sharm, A. Jamwal, A. Aggarwal, S. Bhardwaj, R. Sood, Major challenges in adoption of RFID for Indian SME's. Int. Res. J. Eng. Technol **5**, 2247–2255 (2018)
33. R. Khorshidi, A. Hassani, A.H. Rauof, M. Emamy, Selection of an optimal refinement condition to achieve maximum tensile properties of Al–15% Mg_2Si composite based on TOPSIS method. Mater. Des. **46**, 442–450 (2013)
34. P.L. Singh, R. Sindhwani, N.K. Dua, A. Jamwal, A. Aggarwal, A. Iqbal, N. Gautam, Evaluation of common barriers to the combined lean-green-agile manufacturing system by two-way assessment method, in *Advances in Industrial and Production Engineering* (Springer, Singapore, 2019), pp. 653–672
35. P. Senthil, S. Vinodh, A.K. Singh, Parametric optimisation of EDM on Al-Cu/TiB2 in-situ metal matrix composites using TOPSIS method. Int. J. Mach. Mach. Mater. **16**(1), 80–94 (2014)
36. M. Akbari, M.H. Shojaeefard, P. Asadi, A. Khalkhali, Hybrid multi-objective optimization of microstructural and mechanical properties of B4C/A356 composites fabricated by FSP using TOPSIS and modified NSGA-II. Trans. Nonferr. Met. Soc. China **27**(11), 2317–2333 (2017)
37. M. Alemi-Ardakani, A.S. Milani, S. Yannacopoulos, G. Shokouhi, On the effect of subjective, objective and combinative weighting in multiple criteria decision making: a case study on impact optimization of composites. Expert Syst. Appl. **46**, 426–438 (2016)
38. B.C. Routara, R.K. Bhuyan, A.K. Parida, Application of the entropy weight and TOPSIS method on Al–12% SiC Metal Matrix Composite during EDM. Int. J. Manuf. Mater. Mech. Eng. (IJMMME) **4**(4), 49–63 (2014)
39. T. Singh, A. Patnaik, B. Gangil, R. Chauhan, Optimization of tribo-performance of brake friction materials: effect of nano filler. Wear **324**, 10–16 (2015)

Chapter 8
Hybrid Approach for Prediction and Modelling of Abrasive Water Jet Machining Parameter on Al-NiTi Composites

S. Rajesh, Anish Nair, M. Adam Khan, and N. Rajini

Abstract In this work, pure aluminium and NiTi is used as matrix and reinforcement to fabricate a smart composite. To get an improved mechanical property, primarily the powder metallurgy process parameters are optimized, and the best processing parameters are used for the fabrication of composites materials and subsequently the composite is used for machining studies. Abrasive Water Jet Machining (AWJM) process is used to study the machinability characteristics of the Al- NiTi smart composites. To study the effect of AWJM parameters on Al-NiTi composites, the following control variables identified are Transverse Speed (TS), Applied Pressure (AP), Standoff Distance (SoD), % Wt. of reinforcements (wt%), Abrasive Size (AS). The output indices are Surface Roughness (R_a) and Kerf Angle (K_a). The experiments are designed and conducted based on the design of experiment. Further, it describes the effectiveness of the hybrid algorithm in predicting and optimizing the Abrasive Water Jet Machining (AWJM) parameters. Grey Relational Analysis (GRA) is used as a feature selection and optimizing tool. The result of feature selection by GRA–Entropy, reveals that the most influencing control variables are ranked in the order as AS, AP, TS, wt% and SoD. Modelling of AWJM process is done by Support Vector Machine algorithm (SVM), and the performance of the model is compared with SVM hybrid models. A hybrid model is developed with the concept of Differential Evolutionary algorithm (DE) and Entropy. Hybrid SVM–Entropy model displayed increased prediction performance by 37.8% compared to the SVM model. GRA–SVM–Entropy hybrid model is compared with the SVM model, it is found that the prediction performance of the GRA–SVM–Entropy hybrid model increased by

S. Rajesh (✉) · A. Nair · M. Adam Khan · N. Rajini
Department of Mechanical Engineering, Kalasalingam University, Krishnankoil, India
e-mail: s.rajesh@klu.ac.in

A. Nair
e-mail: anish@klu.ac.in

M. Adam Khan
e-mail: adamkhanm@gmail.com

N. Rajini
e-mail: n.rajini@klu.ac.in

© Springer Nature Switzerland AG 2021
S. Pathak (ed.), *Intelligent Manufacturing*, Materials Forming, Machining
and Tribology, https://doi.org/10.1007/978-3-030-50312-3_8

49.1%. It is found from the GRA–Entropy method; the optimal conditions are A2, B1, C1, D3, and E1.

Keywords Al-NiTi · AWJM · GRA · Entropy · SVM · DE

8.1 Introduction

The demands for the advanced materials in modern manufacturing industries (auto-mobile, aerospace and advanced electronics and biomedical system) are in exponential form because of the competitiveness among the industries. In advance materials, the importance of metal matrix composites cannot be alleviated because of its excellent strength to stiffness ratio. Almost in all aerospace and automobile industries, major components are being replaced by metal matrix composites. The uniqueness of this material is strength and stiffness, which enables them to act as load-bearing members in structural applications [1]. Within the metal matrix family, aluminium composites play the lead role owing to the lightweight nature and easiness in the fabrication process. Most commonly used reinforcements are ceramic-based material which is in the form of fibre, flakes and particulates [2]. NiTi could be one of the alternative reinforcement material for ceramics because of its excellent thermal and corrosion resistance properties. High thermal expansion property of the NiTi would bring better toughness and tensile strength to the aluminium composites [3, 4]. There are various pieces of literature supporting and stating that the addition of NiTi on aluminium and its alloys increases the tensile strength, pre-strain on yield strength [5, 6]. The following are the most widely used processes for the fabrication of aluminium-based composites, conventional casting, centrifugal casting, squeeze casting, hot pressing and ultrasonic consolidation [7]. Among the listed, Powder metallurgy (PM) is one of the efficient processes for the fabrication of aluminium-based composites nearer to the net shape even at low temperature [4–8].

However, the secondary machining operation is necessary for the finished powder metallurgy component to bring to the desired dimension and surface finish. The problem with conventional machining operations is an extreme tool wear and imperfect surface finish of the composites. Lin et al. [9] reported that the existence of intermetallic compounds in the NiTi reinforced composites lead to strain hardening and makes the machining process difficult leading to poor surface texture. In recent times, nonconventional machining processes are deemed superior in handling the issue mentioned above with ease and able to produce an excellent surface finish with less tool wear or no wear, the example of such machining processes are Laser Beam Machining, Electric Discharge Machining and Water Jet Machining (WJM) [8, 10–12].

Kong et al. [13] reported that the performance of the AWJM is better than WJM while machining the NiTi materials. Frotscher et al. [14] noted that thermal distortion of NiTi materials is less while machining with AWJM compared with micromachining. Material Removal Rate, Surface finish and Kerf angle of the AWJM process rely on the various control variables such as abrasive parameters (size, shape, diameter, type of abrasive materials, and abrasive mass flow rate), hydraulic parameter (water flow rate, pump pressure, and orifice diameter) and machining parameter (transverse feed, standoff distance, impact angle, focus diameter, and several passes) [9].

Therefore, Surface roughness (R_a) and Kerf Angle (K_a) is the critical indicator in AWJM process. Scientific approaches are used to find the most favourable fusion of the machining parameter using different statistical methods. Researchers have also used different single, multiobjective and soft computing tools to establish the most favourable fusion of the machining parameter. Amidst all other different optimization tools, Grey Relational Analysis (GRA) is a useful methodology to solve the multiobjective optimization problem. The application of GRA is unlimited in solving the machining problems [15, 16]. Though GRA is a useful tool, and it suffers from two problems. The first problem with the discrete optimization approach is, it has relatively high error since computation is done only on the discrete samples, and another is if its weightage for output performance is not correctly assigned it leads to error in deriving the optimal solution.

The problem of the later can be alleviated by suitable weight computational methods like PCA, Entropy, fuzzy and former can be addressed by using suitable intelligent methods. Nair and Kumanan (2017) applied weighted principal component analysis to estimate the weight of the output performance characteristics by considering the AP, SoD, TS and abrasive volume as input parameters [17]. Lu et al. (2009) adopted the GRA–PCA method to optimize the high-end milling parameter; the PCA method is used to compute the weight to estimate the grey relational coefficients. The PCA method is able to predict the exact weight for each performance characteristics, and the optimal solution is derived near to it [16]. Rajesh et al. (2014) employed the GRA–Entropy method to optimize the tuning parameter to derive the near-optimal solution. The advantage of entropy over the PCA method is, it is simple in the procedure. The method, as mentioned above, can address the problem of calculating the exact grey relational coefficient [18]. On the other side, the intelligent method has the ability to identify and generalize robustly, so that it is capable of handling uncertainty and nonlinear problems. Zhou et al. (2017) used RBF neural network model to predict the grey relational coefficient of the ball-end milling process. The researcher has attempted to solve the complexity and uncertainty problem, as it is possible to convert discrete optimization into a continuous optimization problem by ANN, fuzzy, PCA and entropy-based methods [19]. The problem with neural network approaches is the generalization and overfitted model, which in turn affects the training accuracy. Hybridization of the statistical tool along with soft computing tool is tried by many researchers to transform the discrete optimization into continuous optimization process to reduce the error of modelling and to decrease the time of computation [20].

Support vector machine is one of the intelligent methods to solve the complex problem with small size, whereas other intelligent methods like ANN requires large sample size [15, 21, 22]. By considering all the facts from the literature, it is necessary to develop an algorithm which is capable of predicting and modelling the complex

AWJM machining process with greater accuracy and through simple procedures. One of the ways to reduce the computational time is by reducing the number of input layers during the training and test process. In this work, an attempt is made to predict and optimize the AWJM parameters using Grey Relational Analysis coupled with Support vector machine by entropy and differential evolutionary algorithm. In addition to that, the GRA method is also used as a feature selection method to develop a hybrid model.

8.2 Materials and Methods

Particulate form of Aluminum and NiTi is used as the matrix and reinforcement material for the fabrication of varying wt, % of NiTi reinforced aluminum metal matrix composites. Average particulate size of the matrix and reinforcement are in the spectrum of <45 μm, and it is purchased from Oxford Laboratory, Mumbai. The required quantity of matrix and reinforcement materials are blended in a planetary ball mill to ensure the homogenous mixture of NiTi and aluminum particulates. The particulates are compacted to the required shape by applying the pressure of 300 kPa in Universal Testing Machine and followed by a sintering process. Before sintering process, the green compacted specimen are glass sealed and placed in a furnace for one hour, and the temperature is kept 550 °C at nitrogen atmosphere. Microstructural studies were done using Keller's agents per the standard metallographic techniques. JEOL JSM-6480 LV scanning electron microscope is used to observe the microstructure in secondary electron (SE) and backscattered electron (BSE) modes.

8.3 Multi Objective Optimization and Modelling of Abrasive Water Jet Machining Process

For examining the behaviour of AWJM constraints on NiTi reinforced aluminium composites, the following parameters are identified. The foremost persuading control variables (TS, AP, SoD, wt%, and AS) are the indexes for deciding R_a and K_a. The experiments are premeditated based on the design of experiment. AWJM machine supplied by Dardi International Corporations (Model DIPS6-2236) is used to carry out the premeditated experiments. Table 8.1 describes the premeditated control variables along with its ranges, units and abbreviations. R_a and K_a are considered as response characteristics.

MITUTOYO Surf 301 is utilized to compute the R_a; the direction of measurement is taken as upright to the direction of flow. The sampling length of 6 mm and 0.6 mm/s speed is used for the measurement of the R_a in three locations. To maintain and ensure consistency, the experiment is repeated twice, and it is presented in Table 8.2. Kerf variation in the machined surface is measured with the help of Optical Microscope

Table 8.1 Control variables and it's level

Control variables	Units	Symbol	Level		
			1	2	3
Transverse speed	mm/min	TS	20	30	40
Water jet pressure	MPa	AP	150	200	250
Standoff distance	mm	SoD	1	2	3
% of reinforcement	%	wt%	20	30	40
Abrasive size	μm	AS	80	100	120

Table 8.2 Experimental runs and its measured results

Run	TS	AP	SoD	%.wt	AS	R_a	K_a
1	20	150	1	20	80	4.34	0.44
2	20	150	1	20	100	4.68	0.57
3	20	150	1	20	120	5.83	0.88
4	20	200	2	30	80	4.67	0.67
5	20	200	2	30	100	5.87	0.97
6	20	200	2	30	120	6.03	1.21
7	20	250	3	40	80	4.81	0.89
8	20	250	3	40	100	5.97	1.23
9	20	250	3	40	120	6.28	1.67
10	30	150	1	40	80	4.36	0.42
11	30	150	1	40	100	6.02	0.48
12	30	150	1	40	120	6.55	0.66
13	30	200	2	20	80	5.26	0.62
14	30	200	2	20	100	7.01	0.68
15	30	200	2	20	120	7.99	0.87
16	30	250	3	30	80	6.32	0.76
17	30	250	3	30	100	7.87	1.07
18	30	250	3	30	120	8.43	1.33
19	40	150	1	30	80	6.49	0.41
20	40	150	1	30	100	7.89	0.45
21	40	150	1	30	120	8.57	0.55
22	40	200	2	40	80	6.73	0.43
23	40	200	2	40	100	8.51	0.44
24	40	200	2	40	120	8.88	0.61
25	40	250	3	20	80	6.97	0.51
26	40	250	3	20	100	9.03	0.79
27	40	250	3	20	120	9.55	0.93

(Make; Motic) and the K_a is calculated based on the Eq. (8.1). The width dimension is measured in three places; average K_a is calculated and reported.

$$K_a = \tan^{-1}\left(\frac{W_{Top} - W_{Bottom}}{2t}\right) \tag{8.1}$$

where, K_a is the kerf angle, t is the thickness of the plate and W_{Top} and W_{Bottom} are the top and bottom kerf width.

8.4 Grey Relational Analysis–Entropy Method

Professor J. Deng developed the concept of grey system theory; this theory assists in solving a higher-order complex problem when the data is with more uncertainties, multiple-input and with discrete properties. Information in the higher-order problem is classified as white, black and grey. If the structure information is fully known, it is referred to as a white system, if it is not; it is referred to as the black system. In the grey system, partial information is identified, and partial information is unidentified.

Let X be the AWJM data set, $G(X)$ is a grey set and it is expressed by the following Eq. (8.2).

$$G(X) = \begin{cases} \mu_u(x) : x \to [0, 1] \\ \mu_l(x) : x \to [0, 1] \end{cases} \tag{8.2}$$

where $\mu_u(x) \geq \mu_l(x); x \in X$ and $\mu_u(x)$ and $\mu_l(x)$ are the higher and lower membership functions. In Grey system theory, decision matrix is framed from the set of alternatives. GRA is the crucial method to frame the decision matrix from the set of alternatives. The following steps are implemented to compute the grey relational grade from the decision matrix. (a) Formation of reference series (b) Normalization of data set (c) Calculation of the grey relational coefficient and (d) Calculation of the degree of the grey equation coefficient.

The reference series is derived from the decision matrix. Equation (8.3) represents the decision matrix.

$$X = \begin{bmatrix} x_{11} & x_{12} & \cdots & x_{1n} \\ x_{21} & x_{22} & \cdots & x_{2n} \\ \cdots & \cdots & \cdots\cdots \\ x_{m1} & x_{m2} & \cdots & x_{mn} \end{bmatrix} \tag{8.3}$$

The decision matrix is normalized as $X_i(j)$ where $(0 \leq x_i(j) \leq 1)$ by Eq. (8.4), to elude the consequence of adopting different units and to diminish the inconsistency. The expectation is smaller, the better, then Eq. (8.4) could be used to normalize the experimental data.

$$X_i^*(j) = \frac{\max x_i(j) - x_i(j)}{\max x_i(j) - \min x_i(j)} \tag{8.4}$$

where, $i = 1, 2, \ldots m$ is the experiments $j = 1, 2 \ldots n$ is the output responses.

The absolute difference between the normalized reference series and corresponding experimental series is determined by the following Eq. (8.5).

$$\Delta_i(j) = abs\left(X_o^*(j) - X_i^*(j)\right) \tag{8.5}$$

where $X_o^*(j)$ = reference series value of the jth criterion; $X_i^*(j)$ = normalized value of the jth criterion. The association between the best and actual results can be represented with the term called Grey relational coefficient, and it is estimated by Eq. (8.6).

$$\gamma_i(j) = \frac{\Delta\min + \xi\Delta\max}{\Delta_i(j) + \xi\Delta\max} \tag{8.6}$$

where, $\Delta\min = \min_i \min_j \Delta_i(j)$; $\Delta\max = \max_i \max_j \Delta_i(j)$ and ξ = distinguished coefficient $\xi\varepsilon[0, 1]$. In broad-spectrum, the distinguished coefficient is presumed as 0.5, irrespective of the nature of the problem.

This thumb rule may mislead to the solution when the problem is of higher-order and complex in nature. To alleviate this problem and to precisely predict the distinguished coefficient, Entropy method is adopted, and the following additional steps are considered to find the grey relational coefficient. The entropy E_j of the set of alternatives for criterion j from the normalized matrix is determined by using the Eq. (8.7).

$$E_j = \frac{1}{\ln(m)} \sum_{i=1}^{m} p_{ij}\ln(p_{ij}) \tag{8.7}$$

where, 'm' is the alternatives. Subsequently the degree of diversification is determined by Eq. (8.8).

$$D_j = 1 - E_j \tag{8.8}$$

Finally, the weight of each criterion is calculated by the following Eq. (8.9), and the degree of the grey equation coefficient is calculated by the Eq. (8.10).

$$W_i(j) = \frac{D_j}{\sum_{j=1}^{n} D_j} \tag{8.9}$$

$$\Gamma = \sum_i^{3} \sum_{j=1} \left[W_i(j) \times \gamma_i(j)\right] \tag{8.10}$$

where $W_i(j)$ = weightage of criterion j. $\acute{\Gamma}_i$ represents the level of association between the reference and comparability sequence. If two sequences are indistinguishable, then the importance of the $\acute{\Gamma}_i$ equals one. It also indicates the degree of influence executed by the comparability sequence on the reference sequence.

8.5 Support Vector Machine and Differential Evolutionary Algorithm

A supervised batch learning system (Support Vector Machine) concept is used to classify/predict the performance of the system/process. A supervised batch learning system utilizes the power of statistical techniques and machine learning theory. This enables, SVM to handle large dimensional data and effectively predict the performance of the process/system [20]. Any algorithm aims to establish a good prediction model. In the case of a supervised batch leaning system, the prediction performance depends upon the model. For example, if the linear model is proposed to develop the hyperplane from the set of experimental results, the function can be expressed in the form of Eq. (8.11).

$$f(x) = w \cdot x + b \tag{8.11}$$

where w, b and x are the weight factors, b is the bias term, and x is the multivariate control variable (input parameters). If the nonlinear model is proposed, the kernel function must be used to map the control variables and performance characteristics by transforming the data in the dimensional space. The performance of the model depends upon the selection of optimal kernel functions and biased conditions. The non-linear model can be expressed as in the Eq. (8.12)

$$R_r(f) = \frac{1}{2}\|w\|^2 + C \sum_{i=1}^{n} L(y_i), f(x_i) \tag{8.12}$$

where $L(y_i)$, $f(x_i)$ is the loss function, and it is used to avoid the problem of overfitting of the experimental data during the training process. There are several loss functions which are developed, and the selection of suitable loss function is the stochastic nature of the problem. In general, for process modelling the most widely used model is ε—insensitive loss function model. The term C is used as a penalization term, to give better trade-off between the complexity of the model and flatness. In Eq. (8.11), the concept of slack variables is introduced to cope with infeasible constraints. Modified Eq. (8.12) is represented as Eq. (8.13), by including the slack variables,

Figure 8.1 shows the methodology adopted to develop a hybrid model for the prediction and modelling of AWJM parameters (Fig. 8.2).

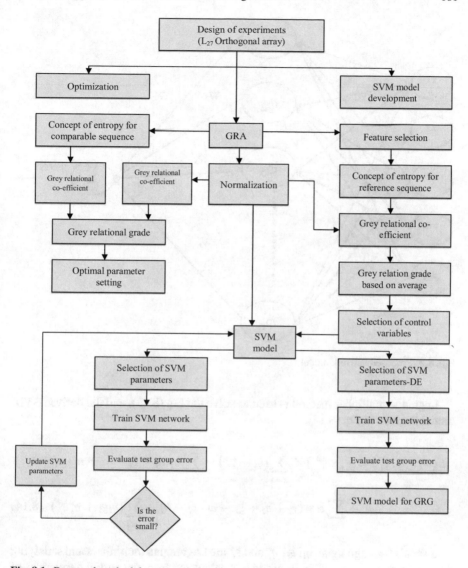

Fig. 8.1 Proposed methodology

$$Minimize = \frac{1}{2}\|w\|^2 + C \sum_{i=1}^{n}(\xi_i - \xi_i^*) \qquad (8.13)$$

Subject to $wx - b - y \leq \xi + \xi_i^*, i = 1, 2, \dots..l$

$$+\xi_i^* \geq 0,$$

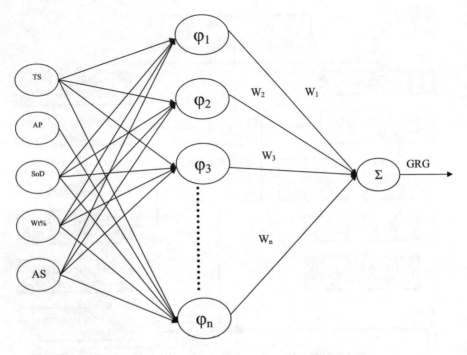

Fig. 8.2 Proposed network model

Lagrange multiplier method is used to solve the Eq. (8.13), and the derived SVM model is given in Eq. (8.14).

$$L(w, b, \xi, \xi_i^*) = \frac{1}{2}\|w\|^2 + C\sum_{i=1}^{n}(\xi_i - \xi_i^*) - \sum_{i=1}^{n}\alpha_i(\varepsilon_i + \xi_i - y_i + w \cdot x_i + b)$$

$$- \sum_{i=1}^{n}\alpha * (y_i + \varepsilon_i + \xi_i - w \cdot x_i - b) - \sum_{i=1}^{n}(\eta_i\xi_i + \eta_i^*\xi_i^*) \quad (8.14)$$

L stands for Lagrangian η_i, ξ_i, η_i^* and ξ_i^* are Lagrangian multipliers and satisfying the condition $\eta_i, \xi_i, \eta_i^*, \xi_i^* \geq 0$, w and b is calculated from the partial derivatives of the L concerning $\eta_i, \xi_i, \eta_i^*, \xi_i^*$.

Gaussian radial basis function with σ standard deviation is commonly used for its better potentiality to handle higher dimensional input space.

$$K(x_i x) = \exp\left(-\|x_i - x\|^2 / 2\sigma^2\right) \quad (8.15)$$

The final model for a better choice of C, ξ and σ can be represented as

$$f(x) = \sum_{i=1}^{n} (\alpha_i - \alpha_i^*) K(x_i, x) + b \tag{16}$$

8.6 Results and Discussion

The first part discusses the optimal parameter setting by GRA traditional, and its hybrid approach and the subsequent section describes performances of SVM model developed by traditional approach and its hybridization.

8.6.1 Optimal Parameter Setting by GRA and GRA–Entropy Method

The correlation between control variables and two performance characteristics is studied by Grey Relational Grade approach since GRG value is the representative of two performance characteristics. The larger value of GRG is the most influencing factor on the two performance characteristics. Table 8.2 shows the comparable sequence and reference sequence used to estimate the GRG. A comparable sequence is the control variables whereas two performance characteristics are considered as the reference sequence (Table 8.3).

Firstly, in the pre-processing stage; the experimental data is pre-processed; the experimental data are transformed into dimensionless number and sequence, by using linear normalization Eq. (8.4). The normalized calculated data for the reference sequence is shown in Table 8.4. The control variable and performance characteristic is normalized between 0 and 1. In Eq. (8.4) $X_i(j)$ represents the maximum value in the respective column and $X_i(j)$ corresponds to the respective experiment. The normalized value of each performance characteristics is used to compute the grey relational coefficient by Eq. (8.6). The deviation sequence Δ_{min}, Δ_{max} and $\Delta_{i(j)}$ are determined using Eq. (8.6). In Eq. (8.6) ε represents the distinguished coefficient, and in general, it is assumed as 0.5. Equation (8.12) is used to compute the grey relational grade. Table 8.4 shows the grey relational coefficient and grey relational grades for R_a and K_a.

From Table 8.4, it is understood that the experiment no. 10 is having higher grey relational grade, and it profoundly influences the optimal result. The average grey relational grade is computed, and it is found that the abrasive size is the most influencing factor on R_a and K_a followed by applied pressure, transverse speed, wt% of reinforcement and standoff distance. Table 8.5 shows the average grey relational grade and optimal parameter setting for the experimental result. It is observed that the optimal parameters are set at A_1, B_1, C_2, D_3, and E_1.

Table 8.3 Normalized data, grey coefficients and grades for comparable sequence

Run	Normalized data		Grey relational coefficient		Grey relational grade
1	1.0000	0.9762	1.0000	0.9696	0.9848
2	0.9347	0.8730	0.7862	0.8568	0.8215
3	0.7140	0.6270	0.4563	0.6708	0.5635
4	0.9367	0.7937	0.7912	0.7865	0.7888
5	0.7063	0.5556	0.4497	0.6310	0.5404
6	0.6756	0.3651	0.4252	0.5448	0.4850
7	0.9098	0.6190	0.7268	0.6661	0.6965
8	0.6871	0.3492	0.4341	0.5387	0.4864
9	0.6276	0.0000	0.3919	0.4318	0.4119
10	0.9962	0.9921	0.9843	0.9897	0.9870
11	0.6775	0.9444	0.4267	0.9319	0.6793
12	0.5758	0.8016	0.3613	0.7930	0.5772
13	0.8234	0.8333	0.5761	0.8201	0.6981
14	0.4875	0.7857	0.3189	0.7801	0.5495
15	0.2994	0.6349	0.2552	0.6755	0.4653
16	0.6200	0.7222	0.3871	0.7323	0.5597
17	0.3225	0.4762	0.2616	0.5920	0.4268
18	0.2150	0.2698	0.2341	0.5100	0.3721
19	0.5873	1.0000	0.3677	1.0000	0.6839
20	0.3186	0.9683	0.2605	0.9599	0.6102
21	0.1881	0.8889	0.2282	0.8724	0.5503
22	0.5413	0.9841	0.3435	0.9795	0.6615
23	0.1996	0.9762	0.2307	0.9696	0.6002
24	0.1286	0.8413	0.2159	0.8272	0.5216
25	0.4952	0.9206	0.3222	0.9054	0.6138
26	0.0998	0.6984	0.2105	0.7159	0.4632
27	0.0000	0.5873	0.1935	0.6481	0.4208

In the traditional approach, the values ξ are assumed as 0.5, but in real-time, the value of the distinguished coefficient may vary based on the selected control variables and performance characteristics. The following steps estimate the exact value of the distinguished coefficient. Equation (8.6) is used to ascertain the association between the control variables and two performances to estimate the ξ. Degree of diversification for the entire set of experimental data is obtained by Eq. (8.8). Based on the degree of diversification and entropy value, weightage for each of the performance characteristics is calculated by Eq. (8.9). It is noted that the weightage for R_a and K_a is 0.76 and 0.24. It is confirmed that the kerf angle plays a vital role in deciding the optimal performance since the weightage for the kerf angle is more than the

Table 8.4 Rank

	TS	AP	SoD	%wt	AS
Level 1	**0.642092**	**0.717515**	0.612408	0.620077	**0.741567**
Level 2	0.590549	0.590049	**0.617277**	0.55746	0.575267
Level 3	0.569495	0.494572	0.572451	**0.6246**	0.485302
Max	0.642092	0.717515	0.617277	0.6246	0.741567
Min	0.569495	0.494572	0.572451	0.55746	0.485302
Max-Min	0.072597	0.222943	0.044826	0.06714	0.256265
Rank	3	2	5	4	1

To highlight the importance of the key values of process and opimization, the reader can understand the optimization process easily

surface roughness. These distinct coefficient values are suitably added in Eq. (8.10) to compute $\acute{\Gamma}_i$ and grey relational grade. The grey relational grade computed by entropy approach is compared with traditional GRA approach, and it is found that there are significant effects of ξ on computing the grey relational grade. However, the ranking predicted by the traditional approach and by entropy approach is the same for the first four rankings, but after that, the order of the ranking is varied. Table 8.6 shows the effect of the distinguished coefficient on grey relational grade and ranking system.

8.7 Performance of SVM and Its Hybrids

In this work, the GRA method is used for two purposes. The first purpose is to select the number of input variables based on their contribution to output performance characteristics, and the other is for establishing the optimum control variables. Optimal setting by traditional GRA and its hybrid is discussed and reported in the first part of the result and discussion. For selecting the most influenced control variable for modelling the SVM, the network is developed by the following procedure. A decision matrix is framed based on Eq. (8.3), and the experimental control variable is normalized using Eq. (8.4). Grey relational coefficient of the control variables is calculated using Eq. (8.6), and it is tabulated in Table 8.7.

From the last row of Table 8.7, it is understood that the average value of normalized data for each control variable and the grey relational coefficient for each control variable is the same. Therefore, it is not possible to find the most influencing control variable on the output performance. To prevail over this, the method suggested by Zhou et al. [15] is used to find the grey relational coefficient. Zhou et al. [15] modified the Eq. (8.6), by considering the effect of each performance characteristics on the control variable to compute the grey relational coefficient. Modification of Eq. (8.6) enables to distinguish the average value of each control variables. Table 8.7 shows the effect of each performance characteristics on the grey relational coefficient, and

Table 8.5 Comparison of grey relational grades and ranks

Ex. No.	Without Entropy weightage		With Entropy weightage	
	Grey relational grade	Rank	Grey relational grade	Rank
1	**0.9773**	**2**	**0.9848**	**2**
2	**0.8410**	**3**	**0.8215**	**3**
3	0.6044	15	0.5635	14
4	0.7977	4	0.7888	4
5	0.5797	18	0.5404	18
6	0.5235	21	0.4850	21
7	0.7074	9	0.6965	6
8	0.5248	20	0.4864	20
9	0.4532	25	0.4119	26
10	**0.9884**	**1**	**0.9870**	**1**
11	0.7540	6	0.6793	8
12	0.6285	13	0.5772	13
13	0.7445	8	0.6981	5
14	0.5969	17	0.5495	17
15	0.4972	22	0.4653	22
16	0.6055	14	0.5597	15
17	0.4565	24	0.4268	24
18	0.3978	27	0.3721	27
19	0.7739	5	0.6839	7
20	0.6818	10	0.6102	11
21	0.5997	16	0.5503	16
22	0.7454	7	0.6615	9
23	0.6695	12	0.6002	12
24	0.5618	19	0.5216	19
25	0.6803	11	0.6138	10
26	0.4904	23	0.4632	23
27	0.4406	26	0.4208	25

To highlight the importance of the key values of process and opimization, the reader can understand the optimization process easily

it is observed that the average grey relational coefficient of each control variable is different, based on the higher average grey relational coefficient value, and ranking is given.

However, Zhou et al. [15] assumed the distinguished coefficient value as 0.5, but the distinguished value coefficient may vary based on the type of process/control variables/output performance as discussed in the first part of the result and discussion. In this work, in addition to the Zhou et al. (2017) suggested method, the concept of

Table 8.6 Normalized value and the grey relational coefficient for control variables

Normalized data					Grey relational coefficient for R_a					Grey relational coefficient for K_a				
TS	AP	SoD	%.wt	AS	TS	AP	SoD	%.wt	AS	TS	AP	SoD	%.wt	AS
1	1	1	1	1	1.00	1.00	1.00	1.00	1.00	0.95	0.95	0.95	0.95	0.95
1	1	1	1	0.5	0.88	0.88	0.88	0.88	0.88	0.80	0.80	0.80	0.80	0.80
1	1	1	1	0	0.64	0.64	0.64	0.64	0.64	0.57	0.57	0.57	0.57	0.57
1	0.5	0.5	0.5	1	0.89	0.89	0.89	0.89	0.89	0.71	0.71	0.71	0.71	0.71
1	0.5	0.5	0.5	0.5	0.63	0.63	0.63	0.63	0.63	0.53	0.53	0.53	0.53	0.53
1	0.5	0.5	0.5	0	0.61	0.61	0.61	0.61	0.61	0.44	0.44	0.44	0.44	0.44
1	0	0	0	1	0.85	0.85	0.85	0.85	0.85	0.57	0.57	0.57	0.57	0.57
1	0	0	0	0.5	0.62	0.62	0.62	0.62	0.62	0.43	0.43	0.43	0.43	0.43
1	0	0	0	0	0.57	0.57	0.57	0.57	0.57	0.33	0.33	0.33	0.33	0.33
0.5	1	1	0	1	0.99	0.99	0.99	0.99	0.99	0.98	0.98	0.98	0.98	0.98
0.5	1	1	0	0.5	0.61	0.61	0.61	0.61	0.61	0.90	0.90	0.90	0.90	0.90
0.5	1	1	0	0	0.54	0.54	0.54	0.54	0.54	0.72	0.72	0.72	0.72	0.72
0.5	0.5	0.5	1	1	0.74	0.74	0.74	0.74	0.74	0.75	0.75	0.75	0.75	0.75
0.5	0.5	0.5	1	0.5	0.49	0.49	0.49	0.49	0.49	0.70	0.70	0.70	0.70	0.70
0.5	0.5	0.5	1	0	0.42	0.42	0.42	0.42	0.42	0.58	0.58	0.58	0.58	0.58
0.5	0	0	0.5	1	0.57	0.57	0.57	0.57	0.57	0.64	0.64	0.64	0.64	0.64
0.5	0	0	0.5	0.5	0.42	0.42	0.42	0.42	0.42	0.49	0.49	0.49	0.49	0.49
0.5	0	0	0.5	0	0.39	0.39	0.39	0.39	0.39	0.41	0.41	0.41	0.41	0.41
0	1	1	0.5	1	0.55	0.55	0.55	0.55	0.55	1.00	1.00	1.00	1.00	1.00
0	1	1	0.5	0.5	0.42	0.42	0.42	0.42	0.42	0.94	0.94	0.94	0.94	0.94
0	1	1	0.5	0	0.38	0.38	0.38	0.38	0.38	0.82	0.82	0.82	0.82	0.82
0	0.5	0.5	0	1	0.52	0.52	0.52	0.52	0.52	0.97	0.97	0.97	0.97	0.97
0	0.5	0.5	0	0.5	0.38	0.38	0.38	0.38	0.38	0.95	0.95	0.95	0.95	0.95
0	0.5	0.5	0	0	0.36	0.36	0.36	0.36	0.36	0.76	0.76	0.76	0.76	0.76
0	0	0	1	1	0.50	0.50	0.50	0.50	0.50	0.86	0.86	0.86	0.86	0.86
0	0	0	1	0.5	0.36	0.36	0.36	0.36	0.36	0.62	0.62	0.62	0.62	0.62
0	0	0	1	0	0.33	0.33	0.33	0.33	0.33	0.55	0.55	0.55	0.55	0.55
0.5	**0.5**	**0.5**	**0.5**	**0.5**	**0.58**	**0.58**	**0.58**	**0.58**	**0.58**	**0.70**	**0.70**	**0.70**	**0.70**	**0.70**

To highlight the importance of the key values of process and opimization, the reader can understand the optimization process easily

entropy is also included in Eq. (8.7) to compute the grey relational coefficient, and it is tabulated in Table 8.7. The average value of the grey relational coefficient computed by both methods is compared and presented in Table 8.7. It is experiential from Table 8.7; the highest grey relational grade is in the order of abrasive size, applied pressure, transverse speed, wt% of reinforcement and standoff distance.

Table 8.7 Grey relational coefficients for the comparable sequence as suggested by Zhou et al. (2017)

Normalized Data					Grey relational coefficient for R_a					Grey relational coefficient for K_a				
TS	AP	SoD	%.wt	AS	TS	AP	SoD	%.wt	AS	TS	AP	SoD	%.wt	AS
1	1	1	1	1	1	1	1	1	1	0.95	0.95	0.95	0.95	0.95
1	1	1	1	0.5	0.88	0.88	0.88	0.88	0.53	0.80	0.80	0.80	0.80	0.57
1	1	1	1	0	0.64	0.64	0.64	0.64	0.41	0.57	0.57	0.57	0.57	0.44
1	0.5	0.5	0.5	1	0.89	0.53	0.53	0.53	0.89	0.71	0.63	0.63	0.63	0.71
1	0.5	0.5	0.5	0.5	0.63	0.71	0.71	0.71	0.71	0.53	0.90	0.90	0.90	0.90
1	0.5	0.5	0.5	0	0.61	0.74	0.74	0.74	0.43	0.44	0.79	0.79	0.79	0.58
1	0	0	0	1	0.85	0.35	0.35	0.35	0.85	0.57	0.45	0.45	0.45	0.57
1	0	0	0	0.5	0.62	0.42	0.42	0.42	0.73	0.43	0.59	0.59	0.59	0.77
1	0	0	0	0	0.57	0.44	0.44	0.44	0.44	0.33	1.00	1.00	1.00	1.00
0.5	1	1	0	1	0.50	0.99	0.50	0.33	0.99	0.50	0.98	0.50	0.34	0.98
0.5	1	1	0	0.5	0.74	0.61	0.74	0.42	0.74	0.53	0.90	0.53	0.35	0.53
0.5	1	1	0	0	0.87	0.54	0.87	0.46	0.46	0.62	0.72	0.62	0.38	0.38
0.5	0.5	0.5	1	1	0.61	0.61	0.38	0.74	0.74	0.60	0.60	0.38	0.75	0.75
0.5	0.5	0.5	1	0.5	0.98	0.98	0.51	0,49	0.98	0.64	0.64	0.39	0.70	0.64
0.5	0.5	0.5	1	0	0.71	0.71	0.63	0.42	0.63	0.79	0.79	0.44	0.58	0.44
0.5	0	0	0.5	1	0.81	0.45	0.57	0.81	0.57	0.69	0.41	0.64	0.69	0.64
0.5	0	0	0.5	0.5	0.74	0.61	0.42	0.74	0.74	0.95	0.51	0.49	0.95	0.95
0.5	0	0	0.5	0	0.64	0.70	0.39	0.64	0.70	0.68	0.65	0.41	0.68	0.65
0	1	1	0.5	1	0.46	0.55	0.46	0.85	0.55	0.33	1.00	0.33	0.50	1.00
0	1	1	0.5	0.5	0.61	0.42	0.61	0.73	0.73	0.34	0.94	0.34	0.52	0.52
0	1	1	0.5	0	0.73	0.38	0.73	0.62	0.73	0.36	0.82	0.36	0.56	0.36
0	0.5	0.5	0	1	0.48	0.92	0.52	0.48	0.52	0.34	0.51	0.97	0.34	0.97
0	0.5	0.5	0	0.5	0.71	0.62	0.38	0.71	0.62	0.34	0.51	0.95	0.34	0.51
0	0.5	0.5	0	0	0.80	0.57	0.36	0.80	0.80	0.37	0.59	0.76	0.37	0.37
0	0	0	1	1	0.50	0.50	0.99	0.50	0.50	0.35	0.35	0.54	0.86	0.86
0	0	0	1	0.5	0.83	0.83	0.56	0.36	0.56	0.42	0.42	0.72	0.62	0.72
0	0	0	1	0	1.00	1.00	0.50	0.33	1.00	0.55	0.46	0.85	0.55	0.46
0.5	**0.5**	**0.5**	**0.5**	**0.5**	**0.72**	**0.66**	**0.59**	**0.60**	**0.69**	**0.55**	**0.68**	**0.63**	**0.62**	**0.68**

To highlight the importance of the key values of process and opimization, the reader can understand the optimization process easily

Table 8.8 Comparison of GRA method and GRA–Entropy method

Control variables	Average weight (without entropy)		Average weight (with Entropy)		Rank
Performance	R_a	K_a	R_a	K_a	
TS	0.700905	0.634381	0.5763	0.6345	3
AP	0.703396	0.638403	0.5171	0.7545	2
SoD	0.696312	0.635624	0.4352	0.7029	5
%.wt	0.698863	0.631447	0.4453	0.6996	4
AS	0.702415	0.642518	0.5449	0.7456	1

From Table 8.7, it is rendered that, the concept of entropy significantly influences the grey relational values; however, it does not alter the rank of the average grey relational grade value. Therefore, GRA and GRA–Entropy method can be used as a feature selection method to rank the control variables (Table 8.8).

Based on the outcome of GRA and GRA–Entropy hybrid method, the SVM model is developed as described in Fig. 8.3 and is discussed in the following section. The first SVM model is developed by considering the input layer with five control variables, one hidden layer and one output layer with grey relational grade. The input neurons are transverse speed, applied pressure, standoff distance, wt% of reinforcement and abrasive size where output neuron is assigned with grey relational grade. Figure 8.4 shows the proposed network model for establishing the relationship between the control variables and output performance.

The successive steps are implemented to develop the SVM model: dividing the experimental data for training and testing, determination of kernel function, γ and σ. The experimental data is divided into training, and testing data, the composition of training and testing is maintained as suggested by Ashanira et al. Training data set is used to establish the linear regression function $y = f(x)$. The linear estimation

Fig. 8.3 Effects of gamma and sigma on MSE

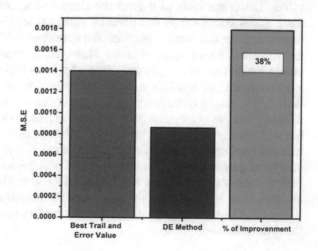

Fig. 8.4 Effects of entropy on MSE

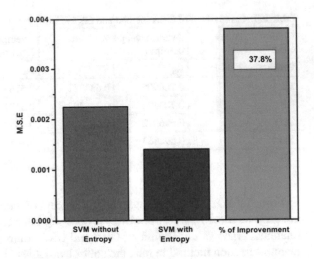

function given in Eq. (8.11) is used to find the unknown parameter w and b. To avoid the fitting problem and to improve the generalization capability, the loss function is also included in Eq. (8.12) by which w^2 is reduced. In SVM statistical learning approach, the accuracy of the model depends upon the selection of proper kernel function and bias values. The values of γ and σ are selected randomly to develop the SVM network model using the training data set, since the selection of kernel function, γ and σ decides the performance of the model. To get the exact value of γ and σ is randomly selected between $10-100$ and $0-1$. Initially, the value of γ and σ is fixed as 10 and 0.2; the performance of the network is analysed. This procedure is repeated by varying the value of γ and σ (100 times with different value of γ and σ), to find the most suitable value to reduce the MSE. The same method is also followed by Ashanira et al. It is observed from the trial and error approach the value of γ near to 10 and σ value near to 0.8 are giving better results compared to other values. Lower the value of γ gives the simplified model, and larger σ indicates a more robust smoothing as described by Agnes et al. The similar kind of result is experienced for this model; therefore, the developed model is capable of predicting the performance with higher accuracy. However, the problem with this method is, it takes a longer time to process and sometimes leads to the faulty selection of γ and σ value. To prevail over this and find the optimal value of γ and σ, DE algorithm is used. The results of both approaches are compared, and it is found that the DE method gives the exact value of γ and σ as similar to the trial and error approach, but with least computational time. Figure 8.5 shows the comparison between the trial and error approach and with the DE method. It is found that the DE method is capable of predicting the optimal of γ and σ with the least computational time and performance of the network is also improved by 38%. The best trial and error value obtained from 100 tests from the trial and error method is compared and presented in Fig. 8.5. Consequently, for the subsequent models, the DE method is used to find the optimal value of γ and σ.

Fig. 8.5 Performance of entropy with varying number of the input layer

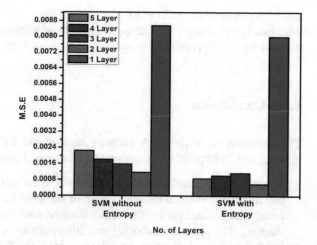

In the second model, the method suggested by Zhou et al. [15] is adopted along with the entropy concept. From Table 8.6, it comprehends that the GRG, with the concept of entropy and without the concept of entropy, is different. Subsequently, the difference in the GRG may improve prediction performance. In this model, the number of input layers and hidden layers are maintained as similar to the first model. The output layer of GRG computed by entropy is considered. The procedure followed for the first model is repeated for the second model, and MSE values are noted and compared. From Fig. 8.4, it is noted that the prediction performance of the entropy-based is 37.8% and is better than the SVM without entropy model.

In the third model, the advantage of the feature selection method is considered to establish the relationship between the input and the output. From Table 8.7, it is understood that the most influencing control variables are abrasive size, applied pressure, transverse speed, wt% of reinforcement and standoff distance. Figure 8.5 shows the comparisons of the ranking of the control variables based on the Zhou et al. (2017) suggested GRA method and GRA and entropy method. It is identified that the ranking of both systems is the same, and the average value of the control variable is slightly different. The value of the GRG is closer to 0.6; accordingly, all the parameters are having a significant contribution to the output as suggested by Ashanira et al. In this model, the number of the input layer is reduced by four by eliminating the standoff distance (rank 5), and GRG without entropy is considered in the output layer.

Similarly, the output layer GRG with entropy is considered for the development of SVM network. The performance of both networks is analysed and reported. The same procedure is repeated by reducing the number of input layer 3, (rank 4 and 5), 2 (rank 3, 4 and 5), and 1(rank 2, 3, 4, and 5) and considering the GRG with entropy and without entropy at the output layer. The performance of the entire network model is shown in Fig. 8.5. It is noted that the number of an input layer with two control variables and one output GRG with entropy is capable of predicting the GRG value with least MSE. It is also noted that the performance of the two layers without entropy and with entropy is capable of predicting the GRG value with least MSE. However, with two layers along with entropy is capable of improving the performance of the network by 49.1% compared to two layers without entropy.

8.8 Conclusion

The effectiveness of the GRA–Entropy method and SVM and its hybrid model is studied, and findings of the experimental results and performances are as follows.

- Traditional GRA and GRA-Entropy hybrid approach do optimization of AWJM parameter, the optimal setting of both of the approaches are compared, and it is found that optimal parameter setting for first four rankings are the same, then it changes. The optimal machining conditions are set at A_2, B_1, C_1, D_3, and E_1.
- GRA is also used as a feature selection tool, the feature selected by the Zhou et al. [15] method and feature selected by entropy method are compared, and it is experimental that the ranking for both approaches is the same, but the values of GRA varies in each experimental result. The most influencing control variables derived from both the methods are abrasive size, applied pressure, transverse speed, wt% of reinforcement and standoff distance.
- Performance of the SVM and hybrid model is studied; implementation of DE and entropy concept increases the performance of the model by 37.8% than the traditional SVM method. Performance of the feature selection method on the SVM model is studied, and it is found that the number of the layer with two control variables improves the performance by 49.1% when compared to the traditional method.

References

1. S.R. Bakshi, D. Lahiri, A. Agarwal, Carbon nanotube reinforced metal matrix composites—a review. Int. Mater. Rev. **55**(1), 41–64 (2010)
2. D.B. Miracle, Metal matrix composites—from science to technological significance. Compos. Sci. Technol. **65**(15–16), 2526–2540 (2005)
3. G.A. Porter, P.K. Liaw, T.N. Tiegs, K.H. Wu, Fatigue and fracture behavior of nickel-titanium shape-memory alloy reinforced aluminum composites. Mater. Sci. Eng., A **314**(1–2), 186–193 (2001)
4. S.L. Angioni, M. Meo, A. Foreman, Impact damage resistance and damage suppression properties of shape memory alloys in hybrid composites—a review. Smart Mater Struct. **20**(1) (2011)
5. M. Dixit, J.W. Newkirk, R.S. Mishra, Properties of friction stir processes Al 1100-NiTi composite. Scr. Mater. **56**, 541–544 (2007)
6. C.L. Xie, M. Hailat, X. Wu, G.M. Newaz, M. Taya, B. Raju, Development of short fiber reinforced NiTi/Al6061composites. ASME J. Eng. Mater. Technol. **129**, 69–76 (2007)

7. D. San Martín, D.D. Risanti, G. Garces, P.E.J. Rivera Diaz del Castillo, S. van der Zwaag, On the production and properties of novel particulate NiTip/AA2124 metal matrix composites. Mater. Sci. Eng., A **526**, 250–252 (2009)

8. O. Akalin, K.V. Ezirmik, M. Urgen, G.M. Newaz, Wear characteristics of NiTi/Al6061 short fiber metal matrix composite reinforced with SiC particulates. J. Tribol. **132**(4) (2010)

9. H.C. Lin, K.M. Lin, Y.C. Chen, A study on the machining characteristics of TiNi shape memory alloys. J. Mater. Process. Technol. **105**, 327–332 (2000)

10. S. Narendranath, M. Manjaiah, S. Basavarajappa, V.N. Gaitonde, Experimental investigations on performance characteristics in wire electro discharge machining of $Ti_{50}Ni_{42.4}Cu_{7.6}$ shape memory alloy. Proc. Inst. Mech. Eng. B J. Eng. Manuf. **227**(8), 1180–1187 (2013)

11. H. Abdizadeh, M. Ashuri, P.T. Moghadam, A. Nouribahadory, H.R. Baharvandi, Improvement in physical and mechanical properties of aluminum/zircon composites fabricated by powder metallurgy method. Mater. Des. **32**(8), 4417–4423 (2011)

12. T.C. Phokane, K. Gupta, M.K. Gupta, Investigations on surface roughness and tribology of miniature brass gears manufactured by abrasive water jet machining. Proc. IMechE, Part C: J. Mech. Eng. Sci. (Sage) **232**(22), 4193–4202 (2018)

13. M.C. Kong, D. Axinte, W. Voice, Challenges in using waterjet machining of NiTi shape memory alloys: an analysis of controlled depth milling. J. Mater. Process. Technol. **211**(6), 959–971 (2011)

14. M. Frotscher, F. Kahleyss, T. Simon, D. Biermann, G. Eggeler, Achieving small structures in thin NiTi sheets for medical applications with water jet and micro machining: a comparison. J. Mater. Eng. Perform. **15**, 776–782·(2011)

15. J. Zhou, J. Ren, C. Yao, Multi-objective optimization of multi-axis ball-end milling Inconel 718 via grey relational analysis coupled with RBF neural network and PSO algorithm. Measurement **102**, 271–285 (2017)

16. H.S. Lu, C.K. Chang, N.C. Hwang, C.T. Chung, Grey relational analysis coupled with principal component analysis for optimization design of the cutting parameters in high-speed end milling. J. Mater. Process. Technol. **209**(8), 3808–3817 (2009)

17. A. Nair, S. Kumanan, Multi-performance optimization of abrasive water jet machining of Inconel 617 using WPCA. Mater. Manuf. Processes **32**(6), 693–699 (2017)

18. S. Rajesh, S. Rajakarunakaran, R. Sudhkarapandian, Optimization of the red mud–aluminum composite in the turning process by the Grey relational analysis with entropy. J. Compos. Mater. **48**(17), 2097–2105 (2014)

19. A.M. Zain, H. Haron, S. Sharif, Estimation of the minimum machining performance in the abrasive waterjet machining using integrated ANN-SA. Expert Syst. Appl. **38**(7), 8316–8326 (2011)

20. T. Verplancke, S. Van Looy, D. Benoit, S. Vansteelandt, P. Depuydt, F. De Turck, J. Decruyenaere, Support vector machine versus logistic regression modeling for prediction of hospital mortality in critically ill patients with haematological malignancies. BMC Med. Inform. Decis. Mak. **8**(56), 1–8 (2008)

21. R. Venkata Rao, V.D. Kalyankar, Optimization of modern machining processes using advanced optimization techniques: a review. Int. J. Adv. Manuf. Technol. **73**(5–8), 1159–1188 (2014)

22. Ulaş Çaydaş, Sami Ekici, Support vector machines models for R_a prediction in CNC turning of AISI 304 austenitic stainless steel. J. Intell. Manuf. **23**(3), 639–650 (2012)

Correction to: Development of $Ti_{50}Ni_{50-x}Co_x$ (X = 1 and 5 at. %) Shape Memory Alloy and Investigation of Input Process Parameters of Wire Spark Discharge Machining

Hargovind Soni, S. Narendranath, M. R. Ramesh, Dumitru Nedelcu,
P. Madindwa Mashinini, and Anil Kumar

Correction to:
Chapter 4 in: S. Pathak (ed.), *Intelligent Manufacturing,*
Materials Forming, Machining and Tribology,
https://doi.org/10.1007/978-3-030-50312-3_4

The book was inadvertently published with the error in chapter author's name. This information has been updated from "Madindwa Mashinin" to "P. Madindwa Mashinini" in the initially published version of the chapter 4. The chapter and book have been updated with the changes.

The updated version of this chapter can be found at
https://doi.org/10.1007/978-3-030-50312-3_4

© Springer Nature Switzerland AG 2021
S. Pathak (ed.), *Intelligent Manufacturing*, Materials Forming, Machining
and Tribology, https://doi.org/10.1007/978-3-030-50312-3_9

Index

© Springer Nature Switzerland AG 2021
S. Pathak (ed.), *Intelligent Manufacturing*, Materials Forming, Machining
and Tribology, https://doi.org/10.1007/978-3-030-50312-3

Printed in the United States
by Baker & Taylor Publisher Services